Henry Forster Baxter

On Organic Polarity Shewing a Connection to Exist Between Organic Forces and Ordinary Polar Forces

Henry Forster Baxter

On Organic Polarity Shewing a Connection to Exist Between Organic Forces and Ordinary Polar Forces

ISBN/EAN: 9783337328030

Printed in Europe, USA, Canada, Australia, Japan

Cover: Foto ©berggeist007 / pixelio.de

More available books at **www.hansebooks.com**

ON

ORGANIC POLARITY.

SHEWING

A CONNECTION TO EXIST BETWEEN

ORGANIC FORCES

AND

ORDINARY POLAR FORCES.

By H. F. BAXTER, Esq.

MEMBER OF THE ROYAL COLLEGE OF SURGEONS, ENGLAND.

———

LONDON,

JOHN CHURCHILL, NEW BURLINGTON STREET.

MDCCCLX.

PREFACE.

THE following Essay contains the results of an investigation which has occupied my attention for some time, subject, however, to the interruptions necessarily connected with professional life. They have for the most part already appeared in Papers published on various occasions, viz. in the Philosophical Transactions, and the Proceedings of the Royal Society of London, the Philosophical Magazine, the Edinburgh New Philosophical Journal, and the Transactions of the Cambridge Philosophical Society, but they are now collected and published in a more connected form, with a hope that they may lead to future inquiries.

Objections might be raised to the title of this Essay, *Organic Polarity.* The results of these investigations lead me to consider it as the most appropriate. It was not selected as the original title to my Papers, but the conclusions that I have arrived at induce me to believe that we have to deal with *polar* forces, since there can be no doubt that some, if not all, of the organic actions which take place in the living body are accompanied with the manifestation of *electrical* action. I might have entitled it, Researches in Animal Electricity, or Electro-Physiological or Biological Researches, but the absurdities associated with the facts comprised under these titles, such as Animal Magnetism, Mesmerism, &c. would only tend to continue, instead of remove, the confused notions connected with the subject, and also the strong prejudices that are entertained in reference to it.

With regard to the conclusions I have arrived at, I am not aware that any serious attempt has been made by *experimental* evidence to prove them erroneous. The objections that have been raised partake unfortunately too much of a personal character. Opposition has been started, in consequence of having in my first Paper, which was communicated and published in the Transactions of the Royal Society, entitled it an attempt to apply some of the discoveries of FARADAY to Physiology; and as some vagueness seemed at the time to be expressed in this title, and to avoid any misconception, I endeavoured in a note to render my meaning more explicit, and to point out the necessity of being acquainted with FARADAY's views in regard to *current force.* Since then, I have been told that ' FARADAY knows nothing of the subject,' and that ' GROVE's views are the most correct,' with other observations which I refrain from relating. It was not my intention, nor had I the least idea at the time, that it would be made and become the subject of a personal affair; but I felt and considered myself bound to state to whom I had been indebted for any success that might attend my labours; for any one at all conversant with experimental researches will readily acknowledge, that to make any advance in the special subject of his inquiry, it becomes absolutely necessary for him to obtain some clear ideas in regard to the subject of his investigations; and I do not think it can be supposed for a moment, that an individual could, without some preliminary knowledge of the subject under consideration, expect to arrive at any definite and trustworthy results. In short, to make any progress in any one particular branch of science would almost appear to depend necessarily upon a previous advanced state of another corresponding branch. I do not suppose any one will deny this; at any rate, it is a conviction that experience has forced upon my mind. I was not so well acquainted with MR. GROVE's views when my Paper was published as at the present time; and, without wishing to derogate from his well-earned reputation as an experimentalist, I have no hesitation in saying, that subsequent knowledge of the subject has only convinced me more than ever of the justness of my first opinion; and I still believe that FARADAY's views will be found to be by far the most philosophic.

I may perhaps have omitted to notice the results obtained by other investigators, which ought to have been alluded to; if so, I have to apologize for the omission; it has been unintentional. I have not attempted to write a Treatise on the subject.

My chief object has been to bring prominently forward those facts which appear to me to be supported by *experimental* evidence, and of paramount importance; the application of these facts must be a subject for future consideration.

I may observe, in conclusion, that whatever success may have attended my labours, I cannot lay claim to any originality; for I shall have the pleasure of shewing, that our predecessors, with all our boasted advantages of superior means of investigation, entertained very correct notions upon the subject. They pursued and exhausted the subject as far as they could, and then threw out conjectures for others to verify.

CAMBRIDGE,
September, 1860.

CONTENTS.

ORGANIC POLARITY.

CHAP. I.

HISTORICAL SKETCH: WALSH; GALVANI; VOLTA; HUM-
BOLDT; VASSALI EANDI; WOLLASTON; NOBILI; MAT-
TEUCCI; DU BOIS REYMOND.

IT is not my intention to enter upon a critical
review of the various opinions that have been entertained in regard to the developement of electrical
action in living animals, nor of the facts bearing
upon the question of the influence of electricity
upon organic actions; but I will endeavour to give
a slight historical sketch of those fundamental facts
which may be adduced as evidence of the manifestation of electrical action in the animal body.
It will, however, frequently happen, that some
difficulty may be experienced in determining this
point, as may be seen in the controversy that
occurred at the time of GALVANI and VOLTA[a].

ᵃ The following works may be referred to.
Traité de l'Electricité et du Magnétism, 7 vols. par M. BECQUEREL.
Paris.
Traité des Phénomènes Electro-Physiologiques des Animaux,
par M. MATTEUCCI, Paris, 1844; and to his Papers published-

That certain fish have the power of giving a *shock* to the human body has been long known, and even employed by the natives of the country in which they are found for remedial purposes ; just as the physicians in the middle of the eighteenth century were in the habit of employing the shocks from the electrical machine in the cure of disease. The shock from the fish and that from the Leyden jar led philosophers to entertain the belief, that they were identical as to their nature, and due to the same power, viz. *Electricity;* whence arose the term, *Animal Electricity.*

In 1773, WALSH[b], in his experiments on the Torpedo, proved to a certain extent this identity : he obtained the *spark,* but could not get any evidence of *attraction* and *repulsion.* He compared the action

since that period in the Philosophical Transactions of the Royal Society of London, and in the various continental scientific Journals.

Du Bois Reymond's large work in German, an abstract of which has appeared in English, edited by Dr. Bence Jones, entitled, *Animal Electricity,* London, Churchill, 1852; and to a Review of the same by Prof. Tyndall, in the British and Foreign Medical Review, Jan. 1854, p. 141.

Rapport sur les Mémoires rélatifs aux Phénomènes Electro-Physiologiques présentés à l'Académie, par M. E. Du Bois Reymond. Comptes Rendus, vol. xxxi. p. 28.

De la Rive's Treatise on Electricity, translated by Walker, 3 vols. Longman.

A Paper published in the Edinburgh New Philosophical Journal, Oct. 1855, entitled, *A brief Review of the Present State of Organic Electricity.* By Prof. Goodsir, F.R.S.L. and E.

[b] Philosophical Transactions, 1773.

of the electric organ to that of a Leyden jar.
" The electricity of the Torpedo," he says, " is
entirely due to the batteries; their upper and under
surfaces are capable, from a state of electric equilibrium, of being instantly thrown, *by a mere energy*,
into a *plus* and *minus* state, like that of a charged
phial, and the current results from a conducting
medium between their opposite surfaces being sup-.
plied naturally or artificially."

Since that period, several philosophers have undertaken the examination of the electric power of the
fish, and the identity between these two powers,
that of the fish and of ordinary electricity, has been
satisfactorily proved ; *attraction* and *repulsion* have
been obtained, the *galvanometer* has been affected,
chemical decomposition produced, and *magnets* made ;
but even at this early period, some of the most
important facts connected with animal electricity
were well known ; and, without intending to de-
preciate the labours of GALVANI, men's minds were
well prepared at his time to seize upon any new fact
that might arise.

In 1791, GALVANI published his celebrated Com-
mentary, entitled, *De Viribus Electricitatis in motu
musculari Commentarius;* but prior to that he was
evidently well acquainted with the circumstance,
that the discharge from an electrical machine would
produce convulsions in a frog. This discovery has
been attributed to chance ; to the circumstance, that
some frogs had been prepared for making broth for

his wife, and were accidentally lying within the influence of an electrical machine when in action.
Upon reading over the various accounts it appears
to me, that the accidental fact observed at the time,
and one of the most importance, was the following:
one of his assistants happened to touch the crural
nerves of a frog with the point of a scalpel, when
contractions of the muscles of the limbs occurred,
(a fact well known to physiologists of the present
day.) GALVANI intuitively saw that effects were now
obtained by the mere point of the scalpel, similar to
those that had been previously observed when a
spark or a shock from the conductor of an electrical
machine traversed the nerve of the animal. Here
was a *new cause* producing muscular contractions, a
new source of power. It was no longer necessary to
have the electrical machine, the mere point of the
scalpel was sufficient.

From our present position we may readily conceive the train of ideas and questions that passed
through GALVANI's mind. A *new fact* was to be
accounted for. Why, he might have asked, should
not the *cause* exist *in* the animal body, for we have
the developement of the power in the fish, as indicated by the *shock*, which is similar to that of a
charged Leyden jar, and this shock *depends* upon
the will of the animal.

The experiments of GALVANI were repeated in
various ways, and by other experimentalists, and
amongst them VOLTA stands preeminent. The *facts*

were not denied; the dispute arose as to the *explanation* of them, and the controversy that succeeded is one of extreme interest. Amongst other important results, the following fundamental facts were established. One by GALVANI, who shewed that when a circuit is formed by means of *any* conducting substance, whether metals or not, between the nerve and the muscle, muscular contractions are produced in the muscles supplied by that nerve, and that the contractions may be produced even *when the free end of the nerve alone touches the external surface of the muscle.* The other by VOLTA, who shewed that the contractions were the *greatest* when two heterogeneous metals were employed to complete the circuit, and that contractions would also be produced when a portion of the nerve alone was included within the circuit. GALVANI explained the effects, by supposing that the muscles acted as reservoirs of electricity, similar to a charged Leyden jar, electricity being formed by the brain; (overlooking the possibility, pardonable at his time, that other parts as well as the brain might produce electricity;) and that the inner and outer surfaces of the muscles being in opposite electric states, when the circuit was completed the contraction of the muscle ensued in consequence of the reunion of the electricity of the two surfaces. VOLTA, on the other hand, considered that the effect was due to the electricity which was developed by the mere *contact* of the heterogeneous metals, or other sub-

stances by means of which the circuit was completed.
Both GALVANI and VOLTA considered, however, that
the *effect*, the *contraction* of the muscle, was evidence
of the manifestation of electrical excitement; in
fact, the limb was then, as it is employed at the
present day, an indicator of electrical action. Both
were wrong in their exclusive notions respecting the
origin of the power; GALVANI in denying or overlooking the circumstance, that the power might
partly arise from the contact of the heterogeneous
substances forming the circuit; VOLTA, on the other
hand, in denying the origin of the power in the
animal body.

HUMBOLDT[c], from his numerous experiments, came
to the conclusion, that the effects could not be entirely referred to the mere. contact of the metals
as suggested by VOLTA, but that they were partly due
to some cause originating in the animal.

I shall now refer to a communication published
at this period, in which will be found a concise view
of the then state of knowledge upon the subject.
It is of some value, inasmuch as it is a letter written
by VASSALI EANDI, at that time one of the celebrated
professors at Turin, to M. DELAMETHRIE, then Secretary to the Royal Academy of Paris, who requested
his opinion " upon galvanism, and the origin of
animal electricity[d]." The position these two

---

[c] Experiences sur le Galvanisme, Paris, 1799.

[d] Lettre de VASSALI EANDI, à J. C. DELAMETHRIE. *Sur le*

individuals held might be adduced as giving some
weight to their authority; and in it will also be found
the details of some experiments, anticipating those
of the present day. " Si on compare," writes VASSALI
EANDI, " le corps animal à la bouteille de Leyde
lorsqu'on approche l'arc conducteur à la boule qui
communique avec l'intérieur de la bouteille tandis
que l'autre extrémité de cet arc en touche la partie
extérieure, on voit les corps légers s'élancer de la
boule à l'arc; le même phénomène devrait avoir lieu
dans la bouteille de Leyde animal, si je puis me
servir de cette expression; cependant, quoique le
D. Valli, le professeur Eandi et plusieurs autres ayent
écrit qu'ils ont observé des mouvements électriques
dans l'expérience de Galvani, comme il s'agit d'une
expérience qui exige la plus grande delicatesse, ce que
la moindre haleine agissant sur les corpuscules
légers, peut tromper l'observateur, je vous dirai
franchement que j'ai répété plusieurs fois cette expérience en changeant l'appareil en faisant usage
de feuilles d'or, et d'autres corps très-legers, et je
n'ai jamais pu m'assurer qu'il en résultat des mouvements électriques." (We here see, that the *evidence*
requisite at that time to confirm GALVANI's conjecture
did not exist, viz. the *attraction* and *repulsion* of light
bodies; the fact being, that the force does not appear
to be of sufficient *intensity* to produce these effects.
The only evidence that could be adduced was the

galvanisme et sur l'origine de l'électricité animale. Journal de
Physique, t. xlviii. p. 336, 1799. Germinal, an. vii.

effect upon the galvanoscopic limb, viz. when the
nerve was laid over on to the surface of the muscles
of the same limb, and which were supplied by the
same nerve, and thus producing contractions; but
this could not be deemed satisfactory, without begging
the question; the existence of the power in the fish
was of some importance, as confirming the conjecture.) VASSALI EANDI goes on to state; " Si j'avais
une opinion à émettre, je serais porté à croire que les
contractions musculaires sont produites par le mouve-
ment de l'électricité animale dirigée par les con-
ducteurs de l'électricité naturelle; car, sans alléguer
en preuve de cette opinion les faits innombrables
publiées par les D. Gardini, Bertholon, Cotugno,
Galvani, Aldini, Valli, Eandi, Giulio, Rossi, Volta,
&c. j'observerai seulement que dans la nature, chaque
corps changeant son état chimique, change aussi sa
capacité propre à contenir le fluide électrique, et
même bien souvent il change de nature par rapport
à l'électricité comme on le voit dans les oxides
métalliques. Or puisqu'il n'y a aucun doute que
l'air dans la respiration, et les alimens, dans la
digestion, ne changent d'état chimique, ils changeront
donc aussi de capacité pour le fluide électrique.
Réad a démontré que l'air dans la respiration, perd
son électricité naturelle; j'ai prouvé ailleurs que les
urines donnent une électricité négative et j'ai fait
voir plusieurs fois aux D. Gerri, Garetti, et aux élèves
de médecine et de chirurgie, que le sang tiré des
veines, donne dans mon appareil électromètrique

(décrit dans le vol. v. de l'Académie des Sciences de Turin, Dec. 19, 1790,) une électricité positive; donc l'électricité naturelle de l'air et des aliments reste dans certaines parties du corps en abondance, tandis que dans le même corps il y a d'autres parties qui, n'en ont pas la quantité proportionnée à leur capacité."

We shall find that the vast mass of facts accumulated at this period by GALVANI, VOLTA, and other experimentalists, required for their full developement a knowledge of a different order of phenomena than those of ordinary electricity, or such as are associated with *static* electricity. The developement of the phenomena connected with *current* electricity and that from chemical action, together with the means of studying this form of electrical action, namely, the galvanometer, were now required, affording sufficient evidence for shewing, that the different departments of science are so intimately and mutually connected, that any advance in one is necessarily dependent upon a corresponding advance in another.

Shortly after GALVANI's death, VOLTA introduced the pile, the voltaic battery; and as he could now obtain the effects without the intervention of organised substances, he was further confirmed in his opinion as to the origin of the power being independent of the animal body; in short, he denied the existence of *Animal Electricity*. Men's minds were now completely withdrawn from the subject of

Animal Electricity; and the next question that arose was that of the *origin* of the power in the Voltaic circle, whether by a *contact* or by *chemical action.* A fresh subject of controversy was now introduced, and one which has continued up to the present day.

In 1806, DAVY astonished the scientific world by his well-known discovery of the *decomposition* of the alkalies by means of the Voltaic battery, and established the important fact, that *acids* were evolved at one pole and *alkalies* at the other pole of the battery, from whence arose the phrase *polar decomposition.* WOLLASTON immediately seized upon the idea, that animal secretions were effected by the agency of a similar electric power, and published an important Paper containing his views. As I have quoted this Paper in the chapter on the Secretions, I shall only now allude to it. Dr. T. YOUNG[d] also entertained views somewhat similar to those of WOLLASTON, but being unsupported by direct experimental evidence, they need not be detailed.

OERSTED's celebrated discovery in 1820, that a magnetic needle was influenced by an electric current, led to the introduction of the *galvanometer;* and in 1827, NOBILI, in his endeavours to ascertain whether this new instrument could equal the delicacy of the galvanoscopic frog in its indications of electrical action, established the important fact, that a current of electricity existed in the limbs of the

[d] Medical Literature, p. 111.

frog, being directed from the feet to the sciatic plexus, or towards the head of the animal, and this he called ' *the proper current of the frog.*' SEEBECK had shortly before ascertained the fundamental facts connected with *thermo-electricity;* and NOBILI, strange to say, was led to believe, that the results he had obtained with the limb of the frog were due to the difference of temperature between the nerves and the muscles.   An interesting example, indicating how prone the mind is to refer any unexplained or residual phenomenon to a newly-discovered class of facts.

Since the introduction of the galvanometer, several observers have attempted to ascertain the existence of electric currents in the living animal.   Amongst them MATTEUCCI stands preeminent, as having established several important facts connected with the developement of electrical action in the animal body ; and to DU BOIS REYMOND also science is indebted for the elucidation of several important points connected with our present subject.   As I shall have occasion hereafter to allude to the results obtained by these and other philosophers, they need not now detain us.

CHAP. II.

As the facts connected with the developement of
electrical action in the living body are so intimately
related to those which occur during ordinary
chemical action, I shall in the present chapter
confine my observations chiefly to the developement
of electricity under these circumstances, and speak,
first, of the electric power, or force, as it exists in
the voltaic circle; and, secondly, of the power as it
exists during ordinary chemical action—during the
reaction of different chemical solutions upon each
other.

First: *Of the force as it exists in the voltaic circle.*
The form which electricity assumes in the voltaic
circle, and termed *current force*, is an indisputable
fact, granted both by the *contact* as well as by the
chemical theorists. As the mass of evidence in
favour of the *chemical origin* of the power in the
circle preponderates over that of the *contact* theory,

I shall confine my remarks to the former, and state what the chemical theory is [a].

'The chemical theory,' says FARADAY, 'assumes, that at the place of action the particles which are in contact act chemically upon each other, and are able under the circumstances to throw more or less of the acting force into dynamic force; that in the most favourable circumstances the whole is converted into dynamic force; that then the amount of current force produced is an exact equivalent of the original chemical force employed; and that in no case (in the voltaic pile) can any electric current be produced, without the active exertion and consumption of an equal amount of chemical force ending in a given amount of chemical change.'

Let us now take a simple voltaic circle, and study the changes that occur in it. I shall assume my readers to be acquainted with the origin of the terms, 'electrode,' 'anode,' and 'cathode [b].' If we take a vessel containing dilute sulphuric acid, and dip two rods, one of zinc and the other of platinum, into it, so long as they remain separate no action occurs on the platinum, but gas is evolved on the zinc, where chemical action is taking place. Let the two rods be brought into communication by means of a conducting substance, such as a metallic wire,

[a] I cannot do better than refer my readers to Faraday's Paper, *On the source of the power in the Voltaic pile*. Experimental Researches, vol. ii. p. 18.

[b] Ibid. vol. i. p. 195.

and the gas is now evolved on the surface of the platinum rod, but the chemical action still continues upon the zinc; in addition to this, we have a current of electricity, as it is called, in the wire, as may be shewn by suspending a magnetic needle parallel to the wire, when the needle will place itself at right angles to it. The explanation of these actions is the following: the oxygen of the water (the electrolyte) combines chemically with the zinc, and the hydrogen, the other element of the electrolyte, is evolved upon the surface of the platinum; during these actions electricity is evolved, and the circuit being completed, it is thrown into dynamic force, *current force*. The water is the decomposing electrolyte; its *anion*, the oxygen, combines at the *anode* with the zinc; whilst the *cation*, the hydrogen, is evolved on the surface of the *cathode*, the platinum. The sulphuric acid, by dissolving the oxide of zinc as it is formed, removes it from the surface of the zinc, and thus assists in continuing the chemical action[c].

Now the *direction* of the current is invariably associated with the direction in which the active chemical forces oblige the *anions* and *cations* to move in the circuit. In the circle we have been speaking of, the action was at the *anode*, at the zinc

[c] A different view may be taken in regard to this explanation of the action of the sulphuric acid. See Daniell and Miller, on the nature of *Compound Electrolytes*. Philosophical Transactions, 1839, p. 97.

surface, and the *direction* of the current was in the direction the *cation* hydrogen may be supposed to move, viz. *from* the zinc across the fluid *to* the platinum, the platinum being the *positive* electrode. I may just observe, that in the majority of instances the *exciting* action is at the *anode*, and the metal acted upon is considered as *positive* to the other metal; but when we speak of the *positive* extremity of the current, this is generally in contact with the metal not acted upon. It is as well to bear this in mind, otherwise it will lead to considerable confusion of ideas, when speaking of the *positive* electrode and *positive metal.*

In the following arrangement, if we take peroxide of manganese, sulphuret of potassium in solution as the electrolyte, and platinum, and form them into a circuit, the platinum will be *negative* to the peroxide of manganese; the current will be in the reverse direction to that observed in the former arrangement. Now whether we suppose the *cation* potassium to abstract the *anion* oxygen from the peroxide or the latter, the *cation* potassium from the sulphur, still the current passes in the direction that the *cation* the potassium may be supposed to move in the circle, and in this case the peroxide will not only be the *positive electrode*, but will also form the so-called *positive metal* of the circuit.

In the following circuit, zinc, muriatic acid, and peroxide of lead, the *acting* points, are at both extremities, at the *anode* and *cathode*. In this instance the two acting points assist each, and so produce a powerful current; the platinum will form the *positive* electrode. But here it may be asked, which forms the *positive* metal?

In all these instances we must not overlook the circumstance of the *decomposition* of the electrolyte; we have a *combination* and a *decomposition* occurring at the same time; which of the two may be considered as the most important act in the developement of the force may perhaps be a question rather difficult to decide. I shall now proceed to speak,

Secondly, *of the developement of current force during ordinary chemical action, during the reactions of different chemical solutions upon each other.* If we take a solution of nitric acid and of potassa, place them on either side of a diaphragm of bladder in a divided cell, and into each cell place a platinum electrode and complete the circuit, the electrode in contact with the acid is positive to the other in the alkali. In the former instances in the voltaic circle, the electrode in contact with the *alkali*, the *cation* was positive to the other. With the solutions the current may be supposed to arise from the *combination* of the acid *with* the alkali; in the voltaic circle to arise from the *decomposition* of the compound from the *separation* of the alkali *from* the acid.

According to BECQUEREL, nitric acid was found to be *positive* with the hydrocholic, acetic, nitrous acids, alkaline solutions, solutions of the nitrates, sulphates, muriates, &c. It was *negative* with sulphuric and phosphoric acids.

Phosphoric acid was *positive* with muriatic, sulphuric, nitric acids; the alkaline and saline solutions.

Water in combining with an acid acts as an alkali; with alkalies as an acid.

With neutral solutions, the most saturated acted as an acid to the less saturated. BECQUEREL considers, that in the reaction of one solution upon another, that which performs the part of an acid takes *positive*, that of an alkali *negative* electricity. The effect, it may be observed, that is produced under these circumstances upon the galvanometer is nothing like so great as that which occurs with the voltaic circle, where metals are employed.

Before quitting the subject of the developement of electricity during chemical action, I shall take the opportunity of speaking of what are called *differential* currents and *combined* currents. If we take a vessel or glass cell and divide it into three compartments by a membrane, so as to form three distinct cells, and into the extreme compartments pour a solution of potash, and into the centre one nitric acid, then place the platinum electrodes of a galvanometer into the potash cells, no effect will occur upon the needle, provided the arrangement be perfect; but if one of the electrodes be placed in the acid, and the other in the alkali, we then have the ordinary result. In the former instance the currents produced, passing in contrary directions and being equal to each other, counteracted each other's effects, and consequently there was no action upon the needle.

Let the cells be filled thus, acid | alkali | acid | alkali, and place the electrodes in the extreme cells; we have here three acting points in the circuit: at one point the current is in one direction, and in two others in a contrary direction; and the action upon the needle in this instance will shew the *differential* result.

The same observations may be made in regard to a voltaic battery, by altering the arrangements of the metals, so as to have a combined action in one case and an interfering action in the other.

I have entered somewhat fully into the developement of electric action during *chemical* action, in consequence of its immediate bearing upon the question of the developement of electric force during *organic* action; and I ought now perhaps to allude to *thermo-electric* action, and to the facts connected with the so-called *catalytic* action or the combining power of platinum; but to enter fully into these questions would occupy too much space, and I must therefore refer my readers to the published treatises on Electricity for further information upon these subjects.

Let us now call to mind the peculiar characters, without alluding to the origin of the power, whether from heat, chemical action, friction, &c. which are asso-

ciated with electric force when manifested in its
static, or in its *dynamic* condition, the former as it is
manifested in a charged Leyden jar, the latter as it
exists in a wire or any conducting medium, when it
assumes the so-called *current* form.   In both instances, in the *static* and the *dynamic* condition, we
have the characters of polarity clearly manifested;
a ' *duality*,' an ' *antithetical action*,' a ' *contrast of
properties in a contrast of positions*;' and it is somewhat surprising to find Mr. GROVE, in 1855, at page
152 of his Essay, stating, that " it is difficult to
convey by words a definite idea of the dual or
antithetic character of force involved in the term
polarity," after the clear definition of it given by
FARADAY, as far back as 1833, when he speaks of it
" *as an axis of power having contrary forces exactly
equal in amount in contrary directions*[d]."   In WHEWELL's
Philosophy of the Inductive Sciences, published in
1840, will be found a very interesting chapter, (Book
V. on the Idea of Polarity,) in which extremely
valuable observations are made in reference to the
same point.   However important Mr. GROVE's views,
as published in his Essay, may be considered, and
that they possess some value cannot be denied, I
still believe that the views advanced by FARADAY,
and worked out in his admirable Memoirs, are by
far the most philosophic.   There is a unity of
thought, a firmness of opinion, and a consistency
about them which bear the stamp of truth.   And

[d] Experimental Researches, vol. i. p. 148.

I cannot refrain from observing, that if the views I have entertained in regard to FARADAY's researches are so erroneous and inconsistent when compared with Mr. GROVE's, I really think that some attempt might and ought by this time to have been made to prove them as such.

It will not be necessary for me to enter into the subject of the construction of the galvanometer, or of the principles of its action; but merely to observe, that for the purpose of electro-chemical research, a galvanometer consisting of several coils of thin wire will be found to be the most appropriate; whilst for the study of thermo-electric and of magneto-electric researches, a galvanometer consisting of a few coils of thick wire or of a plate of copper will be found to be the best. For our present purpose, the former with several coils will be found the most applicable; at the same time we must not overlook the importance of occasionally using the latter.

The mode usually adopted, and found mos convenient to pursue in these experiments, was as follows. The galvanometer was placed and well secured on a firm table or stand, free from any vibrations of the room, and with a good light falling upon it, at the same time not exposed to the direct rays of the sun, so as to avoid the production of air-currents within the shade of the instrument. Two thick copper wires, about six inches in length, were bent at right angles, so that one extremity should be attached to the instrument by means of the binding screws, whilst the other extremities were dipped into two mercurial cups; by these means the galvanometer would remain fixed, and it was at these mercurial cups that the contacts were made and broken. The electrodes, by means of which the circuits were formed and completed in the subject during an experiment, consisted of thick platinum wires about nine inches in length.

There are several circumstances of paramount importance to be attended to during an experiment.

First. The arrangement of the galvanometer should be perfect, and in a good working condition; that is to say, the contacts between the different parts should be complete, and no effects should occur upon the needle when a circuit is formed between the mercurial cups: at the same time, the instrument should indicate any slight action from some well-known cause, and if delicate, the mere squeezing of the extremities of the platinum electrodes between the fingers would be quite sufficient to produce an effect upon the needle.

Secondly. Avoid having any steel about the person, such as pen-knives, metallic buttons, or covered steel buttons in the sleeve of the coat, and when scalpels are used, the experiment should be commenced on a moveable piece of board away from the galvanometer, and then brought to it. As far as possible, the galvanometer should be out of the influence of iron bars, such as those connected with shutters and fire-irons.

Thirdly. Cleanliness is of the utmost importance.
Many experiments have ended in fallacious results
from the mere want of attention to this circumstance.
The platinum electrodes should be perfectly clean,
especially the *extremities;* they should be well rubbed
with a dry clean cloth, after and previous to the
formation of each circuit during the experiment,
and occasionally exposed to the heat of a clear fire.
The secondary current, due to the polarized state of
the platinum electrodes, must not be forgotten; to
remedy this, the ends of the electrodes should be
dipped into water, and their free extremities brought
into contact so as to form a circuit, and then
cleaned. No acids or alkalies should be employed,
or allowed to come into contact with the hands, or
fingers, or the towels, unless for some purpose in
the experiments: the hands and fingers should be
perfectly clean, and no dirty towels used.

There is one observation to be made in reference
to the employment of the galvanometer. It may be
supposed that a great difficulty is experienced in
obtaining any result with this instrument, and that
it is absolutely necessary to employ a *delicate* instrument for this purpose; nothing is more fallacious.
There is no difficulty in getting an effect upon the
needle in these experiments, but the contrary; the
difficulty is not to get an effect. In saying this,
however, it is not to be supposed that no satisfactory
results are to be obtained; I only wish to observe,
that it is not the mere motion of the needle and its

greatest amount that are to be observed as indicating the most satisfactory results, but the constant, steady, and definite motion of the needle; and with respect to the amount of its deflection, a slight rather than one of great extent is that upon which the most trustworthy and satisfactory results are to be placed. Practice alone, however, can only and would soon give that knowledge which cannot be obtained by mere description, and it is by practice alone that any experiment can be safely undertaken and depended upon. We must bear this most important fact also in mind, that it is not the mere circumstance of getting an effect upon the needle that we are to be contented with, but we have to account for the effects when obtained, and to shew to what class of phenomena they may be referred.

CHAP. IV.

ON THE MANIFESTATION OF CURRENT FORCE DURING THE
ORGANIC PROCESS OF SECRETION IN THE LIVING OR
RECENTLY-KILLED ANIMAL; IN THE STOMACH AND
INTESTINES; THE LIVER; THE KIDNEY; THE MAMMARY
GLAND; THE LUNGS.

After DAVY's celebrated discovery, in 1806, of the
decomposition of the alkaline salts by voltaic electricity, and when he had established the important
fact that *acids* were evolved at one pole and *alkalies*
at the other pole of the battery, (from whence arose
the phrase *polar decomposition*,) WOLLASTON immediately seized upon the idea, that the animal secretions were effected by the agency of a power
similar to that of a voltaic circle; and in the Paper[a]
containing this remarkable conjecture, which was
published in 1809, he also suggested, that "the
qualities of each secreted fluid may hereafter instruct
us as to the species of electricity that prevails in
each organ of the body;" that as the stomach and
kidneys secreted an acid, for example, whilst the
liver secreted an alkaline compound, the two former
might indicate a *positive* electric state or condition,
and the latter a *negative* state or condition. PROUT[b]
cautiously advanced a somewhat similar opinion, and
says, "Admitting that the decomposition of the salt

[a] *Philosophical Magazine*, vol. xxxiii. p. 488.
[b] *On Stomach and Urinary Diseases*, 3rd edit. p. xxv.

of the blood, &c. is owing to the immediate agency of a modification of electricity, we have in the principal digestive organs a kind of galvanic apparatus, of which the mucous membrane of the stomach and intestinal canal, generally, may be considered as the acid or positive pole, while the hepatic system may, on the same view, be considered as the alkaline or negative pole. He also quotes an experiment of MATTEUCCI as, in some degree, confirming his opinion.

DONNE[c], upon applying one of the electrodes of a galvanometer to the stomach and the other to the liver, obtained an effect upon the needle, and the result of this experiment was subsequently confirmed by MATTEUOCI[d].

The suggestion thus thrown out, that the stomach and liver formed *poles* similar to those of a galvanic pile, having apparently received some confirmation from *experimental evidence*, it now became of some importance to trace out the *circuit*, the *path* of the current; and, if possible, the *origin* of the power, so as to complete the whole evidence necessary for the proof of the truth of the suggestion.

Reasoning upon these facts, and assuming that the stomach and liver *did* actually form the two *poles* similar to those of a galvanic circle, it was reasonable to suppose that the *electric current* would pass *from* the stomach *to* the liver by the blood in the portal vein. To

[c] BECQUEREL, *Traité de l'Electricité*, tom. i. p. 327.
[d] Ibid. tom. iv. p. 300.

ascertain the truth of this supposition, I now inserted the two platinum extremities of the electrodes of a galvanometer into the portal vein, and as far apart as possible, in order to obtain the supposed *diverted current;* but no effect was observed. The electrodes were then inserted, one into the *portal vein,* the other into the *hepatic vein,* still no effect.

POUILLET[e] and MULLER[f], it may be observed, had previously ascertained that no effect occurred when they inserted one electrode into an *artery,* and the other into a *vein,* of a living animal.

No evidence could be obtained from these experiments indicative of the *path* of the current; the galvanic circle was therefore not complete; and some of the essential conditions were evidently wanting.

Repeating the experiments of MATTEUCCI upon other animals than rabbits, the effects observed by MATTEUCCI were not always obtained; as these results will again come under consideration, they need not now detain us.

Pondering over these failures, it soon became evident that more correct notions in regard to the *origin* of the power in the voltaic circle were requisite; the term *current* also, with its ordinary associations, (of something flowing in *one* direction,) was a source of great embarrassment; and it was

---

[e] *Journal de Physiologie,* tom. v. p. 5.

[f] MULLER's *Physiology,* translated by BALY, vol. i. p. 148. 2d edit.

thus found that a deeper insight into a knowledge of
FARADAY's opinions in respect both to the *origin* of
the power in the voltaic circle, and to that of *current
force* in particular, viz. as AN AXIS OF POWER HAVING
CONTRARY FORCES EXACTLY EQUAL IN AMOUNT IN CON-
TRARY DIRECTIONS, was absolutely essential. I have
endeavoured in chap. ii. to give a brief account of
his views on this point; and I may also refer to
Whewell's Philosophy of the Inductive Sciences,
vol. i. p. 331. first edition, to the chapter ' On the
Scientific Application of the Idea of Polarity.'

Dismissing the notion that the stomach and liver
are related to each other in the same manner as the
poles of a galvanic circle are mutually dependent, and
with a more correct knowledge of the *origin* of the
power in the galvanic circle derived from FARADAY's
memoirs, the thought arose that it might be during
the *formation* of the secretions where the changes
were actually going on, that the evidence sought
for could possibly be obtained. How far these
surmises were correct will now be seen.

SECT. I. *On the Manifestation of Current Force during
the formation of the Secretions in the mucous mem-
brance of the alimentary canal, viz. the stomach and
intestines.*

The mode of employing the galvanometer and of
conducting the experiments, together with the pre-
cautions necessary to be observed, have been alluded
to in chap. iii.

Although experiments performed upon the living animal may be considered as affording more satisfactory results, nevertheless, as the results can be obtained when sensibility is destroyed, the following mode may be adopted in preference to the use of chloroform.

Let a few drops of strong prussic acid be dropped on the nose, insensibility is thus quickly produced; or let the animal be pithed, and upon laying open the chest or abdomen, the heart will be found to beat and the circulation to continue. Under these circumstances, if the platinum electrodes of a galvanometer are placed one in contact with the mucous surface of the small or large intestine, the other in contact with the blood in a vein from the same part, a deflection of the needle will be obtained indicating a current through the instrument, the electrode in contact with the blood being *positive* to the other in contact with the mucous surface. If the same experiment be repeated with the mucous membrane of the stomach, the effects may vary. If the stomach be empty, then the electrode in contact with the blood of the vein coming from the same part will also be *positive;* but if there be any food in the stomach and should it contain much acid, then the electrode in contact with its mucous surface will most probably indicate a *positive* state. Now these are the fundamental facts and the results, which are readily obtained with proper precautions, may be thus summed up: *when the electrodes of a galvanometer are brought into contact one with the mucous surface of the intestine in a living or recently-killed animal, and the other with the venous blood from the same part, an effect occurs upon the needle indicating the secreted product and the venous blood to be in opposite electric states.*

The amount of deflection of the needle would vary according to the delicacy of the instrument employed; with an ordinary galvanometer, consisting of but few coils, the deflection was from 3° to 8° or 10°.

When the electrode, instead of being in contact with the *venous* blood, is in contact with the *arterial* blood, or the surface of the mesentery, the effects upon the needle are the same, as far as the *direction* of the current is concerned, but the amount of deflection may not be so great.

Let us now endeavour to explain these results according to known actions, such as the chemical reaction of two fluids upon each other, or to the *heterogeneity of fluids*, as it is sometimes called. If, for example, a glass cell be taken having a porous diaphragm in its middle, such as a piece of membrane, so as to divide it into two cells, and into one compartment we pour an acid solution, and into the other an alkaline solution, and then dip the platinum electrodes of a galvanometer into each of these cells, an effect upon the needle is produced indicating the electrode dipping in the acid solution to be *positive* to the other. These facts, which have been well worked out by BECQUEREL[g], may be enunciated in the following proposition : *During the reaction of two fluids upon each other, that which performs the part of an* acid *takes* positive *electricity, and that of an* alkali, negative *electricity.*

In experiments upon animals, as just related, it was found that the electrode in contact with the *venous* blood was *positive* to the other, excepting when there was much acid in the stomach, and then the electrode in contact with the mucous surface of the stomach was *positive* to the other in contact with the blood.   Now in order to explain these results, under the supposition that they arise from the chemical reactions of the fluids upon each other, it must be supposed, that when the electrode in contact with the venous blood is *positive* to the other, that then the blood acts as an *acid,* and not only so, but *combines* with the substances of fluids in the intestines.   When it is found, however, that the electrode in contact with the stomach is *positive,* then it may be supposed, and rightly so, that the results are due to the chemical reactions which occur in that organ between the acids and other fluids that are there found.   But should we be justified in supposing, that when the electrode in contact with the blood is *positive* to the other in the stomach, the stomach being empty or containing but little acid, that then the blood is acting as an acid ?   Here, as in the intestines, it would be

g *Loc. cit.* vol. ii. p. 77.

necessary to assume, that immediately after the *separation* of the secreted product (the acid) from the blood had taken place, that they then immediately recombined; and not only so, but that the blood, in direct opposition to the well-known fact of its alkaline characters, *must* be acid in order to account for the effects produced. It would therefore appear, that no grounds exist for believing that the results obtained in the living animal can be considered as entirely dependent upon the mere reaction of the heterogeneous fluids upon each other, upon their *combination* for example; and without stopping to adduce more arguments against this supposition, let us now proceed to compare the results with another class of phenomena, viz. with those actions which take place in a voltaic circle where *decomposition* is effected.

It will be better to confine our attention to the actions which take place in the *exciting* cell of a voltaic circle where the power *originates*, and withdraw our minds for the present entirely from the changes which take place in the *decomposing* cell of the battery where *polar decompositions* are effected: the principal object being to ascertain whether, during the *decomposition* of a compound, or during the separation of an acid from an alkali, the same effects are produced upon the galvanometer as occurs during the *combination* of an acid with an alkali.

Let us take an elementary circle, zinc, platinum,

and a dilute solution of muriate of soda, and consider the two metals as forming the terminations
of the electrodes of the galvanometer, one of zinc
and the other of platinum, instead of having two
platinum electrodes as heretofore. When the electrodes are dipped into the solution, the actions
which take place are the following : the muriate of
soda is decomposed by the attraction of the zinc
for the chlorine or muriatic acid, whilst the soda is
evolved on the surface of the platinum; now under
these circumstances the platinum electrode, in contact with the soda, is *positive* to the other, and,
according to common phraseology, the direction of
the current is in the same direction as the *cation*
(the alkaline compound, the soda) is supposed to
travel. Here then is a case of *decomposition*, a
separation of an acid *from* an alkali, effected by
chemical agency, and the electrode in contact with
the *alkali* is *positive* to the other in contact with the
acid ; the effect being contrary to that observed
during the *combination* of an acid with an alkali, as
has been just shewn. Let us now compare the
results which occur in the animal with those which
take place in the voltaic circle. When the electrode
is brought into contact with the venous blood, it is
positive to the other in contact with the secreting
surface of the intestine ; if it be now supposed that
the blood is alkaline, and there is every ground for
so doing, the electrode in contact with the blood is
exactly similar to that in contact with the alkali in

D

the voltaic circle; but instead of the secreted product combining with the other electrode, as the acid does in the voltaic circle, it passes away. In the animal the current may be supposed to be dependent upon the *decomposition*—if I may so term it—of the arterial blood, being as it were separated into its two elements, the secreted product and venous blood, just as the muriate of soda is decomposed and separated into its two elements, muriatic acid and soda.

At present, it may be remarked, that no opinion as to the mode in which the secretions are effected is being given; I am only endeavouring to ascertain now what does occur, and to what class of phenomena these actions, those of secretion, bear the greatest resemblance. This subject will again come under our notice.

Before proceeding to shew that in other organs there exists the same manifestation of current force during secretion, I cannot omit noticing the opinion that WOLLASTON entertained in regard to the question now under consideration, and shall therefore quote his own words: "At the time," says WOLLASTON[b], " when MR. DAVY first communicated to me his important experiments on the separation and transfer of chemical agents by means of the voltaic apparatus, which was in the autumn of 1806, I was forcibly struck with the probability that animal secretions were effected by the agency of a similar electric power; since the existence of this power in some

[b] Philosophical Magazine, vol. xxxiii. p. 488.

animals was fully proved by the phenomena of the
Torpedo and of the Gymnotus Electricus; and since
the universal prevalence of similar powers of lower
intensity in other animals was rendered highly
probable by the extreme suddenness with which
the nervous influence is communicated from one
point of the living system to another.

" And though the separation of chemical agents,
as well as their transfer to a distance, and their
transition through solids and through liquids, which
might be expected to oppose their progress, had not
then been effected but by powerful batteries; yet it
appeared highly probable that the weakest electric
energies might be capable of producing the same
effects, though more slowly in proportion to the
weakness of the power employed.

" I accordingly at that time made an experiment
for the elucidating this hypothesis, and communicated it to MR. DAVY and to others of my friends.
But though it was conclusive with regard to the
sufficiency of very feeble powers, it did not appear
deserving of publication, until I could adduce some
evidence of the actual employment of such means
in the animal economy.

" The experiment was conducted as follows:
I took a piece of glass-tube, about three-quarters
of an inch in diameter and nearly two inches long,
open at both ends, and covered one of them with
a piece of clean bladder. Into this little vessel
I poured some water, in which I had dissolved

$\frac{1}{240}$th of its weight of salt; and after placing it upon
a shilling with the bladder slightly moistened
externally, I bent a wire of zinc, so that while one
extremity rested on the shilling, the other might
be immersed about an inch in the water. By successive examinations of the external surface of the
bladder, I found that even this feeble power occasioned soda to be separated from the water, and
to transude through the substance of the bladder.
The presence of alkali was discernible by the application of reddened litmus-paper after two or three
minutes, and was generally manifested even by the
test of turmeric paper before five minutes had
expired.

" The efficacy of powers," continues WOLLASTON,
" so feeble as are here called into action, tends to
confirm the conjecture, that similar agents may be
instrumental in effecting the various animal secretions which have not yet been otherwise explained."

There is one circumstance connected with WOL-
LASTON's conjecture which must be noticed, viz. the
idea that secretion depended *upon*, or is the *effect* of,
a power similar to that which exists in a voltaic
circle; but it must be borne in mind, that the *origin*
of the power in the voltaic circle was not so completely understood at the time WOLLASTON published
his conjecture as it is at the present day; and
although he himself was an advocate for the opinion
that it depended upon chemical action, it nevertheless required the elucidation that it has subse-

quently received at FARADAY's hands; the fact being, that the chemical action which occurs is the *cause* of the power, or, in other words, the current is a mere manifestation of the chemical action that is taking place.   I shall now pass on to the consideration of the manifestation of current force during secretion in other organs; and first, in the liver.

SECT. II. *On the Manifestation of Current Force during Biliary Secretion.*

If the platinum electrodes of the galvanometer be inserted one into the gall-bladder, and the other into the hepatic vein, or which will be found better still, in consequence of the blood flowing over the intestines, into the *vena cava ascendens* in the chest, we then obtain evidence of the manifestation of current force; the electrode in contact with the blood in the vein being *positive* to the other in contact with the bile in the gall-bladder.   The amount of deflection of the needle varies from 5° to 10°.

When the electrode, instead of being inserted into the hepatic vein or into the *vena cava ascendens*, is inserted into the *vena porta*, the other remaining in the gall-bladder, the former will still indicate a *positive* state; but the effect upon the needle is not so great.

Other circuits were formed; viz. between clots of blood and pieces of liver; between the mucous membrane of the intestines and the gall-bladder; between the blood in the chest and various parts of the abdomen; one electrode was coated with bile, and then both of the electrodes were dipped into the blood in the chest. It was generally found that the electrode in contact with the blood was *positive*, but not always; sometimes vibrations of the needle only occurred, at other times the needle went as far as 80° or 90°, and then stopped. The motions of the needle presented quite a different character to those observed when the bile in the gall-bladder and the blood flowing from the vena cava inferior were formed into a circuit; the latter presented a steady character, they could be depended upon; whereas with the former a greater effect might be produced at first, and it would then cease, or perhaps go in the opposite direction. The following conclusion may therefore be deduced : *when the electrodes of a galvanometer are brought into contact, one with the bile in the gall-bladder, and the other with the blood in the hepatic vein, or vena cava ascendens, an effect occurs upon the needle, indicating the secreted product (the bile) and the blood to be in opposite electric states.*

It may be said, and with apparent justice, that if the actions which occur during secretion be similar to those that take place in the exciting cell of a voltaic battery, as was suggested in the previous Section, the electrode in contact with the *alkaline* bile ought now to indicate a *positive* state.

The force of this objection depends entirely upon the assumption that the *bile* contains a *free alkali*. The researches of chemists, and especially LIEBIG, have however shewn, that with the alkaline bases which exist in the bile, are associated peculiar *organic acids*, such as the *bilic, choleic*, &c. As these acid compounds are easily decomposed, we should not be justified in supposing, from finding a number of indestructible basic elements which exist in the ultimate analysis of the bile, that these basic elements therefore existed as such in the composition of the bile; and although the bile may present an alkaline reaction, this alone would not necessarily indicate the existence of a *free alkali.* It would appear more reasonable to suppose, that these basic elements existed in combination with the destructible *organic acids.* Similar remarks may undoubtedly be made respecting the composition of the blood, but the chemical evidence in favour of the existence of a free alkali in the blood is far stronger than that for its existence in the bile. The opinion that the fluidity of the blood may be dependent upon the alkaline salts has been long entertained by physiologists, and would appear to have received strong confirmation from the recent experiments of DR. RICHARDSON[i], to which I may refer.

Having so far removed this objection, the same remarks that were made in regard to the secretion that occurs in the intestinal canal, and which I need not recapitulate, may now be applied to the formation of the bile. So here in another class of secretions,

---

[i] *The Cause of the Coagulation of the Blood.* Churchill, 1858.

additional evidence has been obtained of the manifestation of current force during secretion.

Before passing on to other secretions, I shall now
notice the fallacy of supposing that the stomach and
liver form poles similar to those of a galvanic battery,
an idea that has been entertained by several individuals. No evidence could be obtained to shew
that the stomach forms the *positive* and the liver the
negative electrode of a circuit similar to those of a
voltaic circle. It may just as well be supposed, that
the lungs and the stomach, or the lungs and the
kidneys, or the liver and the lungs, or the kidneys
and the lungs, are similarly related, if we are to be
guided by the mere circumstance of their relative
connections in regard to the circulation of the blood
through these different organs. Each organ, the
stomach and liver, would appear to have, however,
its own elementary circle, if I may so express it;
but no evidence exists to shew that these two organs
are so mutually related as to form *one* circle. There
is one fact which is of some interest, and deserving
of notice, it is this; the blood, from which the biliary
secretion is formed, has previously undergone some
most important changes during its passage through
the coats of the stomach and intestines, and thus an
important relationship must necessarily exist between these two organs; and the question may
naturally arise, Is not the blood during its passage
through the coats of the stomach and intestines, and
especially by the stomach, thus deprived of most of

the elements of its fixed acids, such as the muriatic acid for example, and so far accounting for the small proportion of these elements that are found in the bile? It must be observed, that I am not now supposing that *all* the acids found in the stomach *must* necessarily come from the blood, for there can be no doubt that some of the acids are formed in that viscus independent of those that are secreted by that organ. But to enter upon this subject would carry us away from our main object, and I shall therefore leave it.

SECT. III. *On the Manifestation of Current Force during Urinary Secretion.*

Upon inserting one of the extremities of the electrodes of the galvanometer into the pelvis of the kidney, and the extremity of the other electrode into the renal vein of the same kidney, an effect upon the needle is produced indicating the electrode in contact with the blood to be *positive* to the other. A difficulty may sometimes arise in obtaining any effect. The amount of deflection of the needle, when obtained, varies from 3° to 5°.

Should we be justified, in this instance, in supposing that the blood is *acid* to the urine, and not only so, but *combines* with the urine, in order to account for the effects observed upon the galvanometer, when a more satisfactory explanation can be adduced, by regarding the effects as being consequent upon the *separation* of the acid product from the blood, as already advanced in the previous Sections with respect to the other secretions?

The amount of deviation of the needle being small, may be referred to the same causes as were observed to exist with regard to the acid secretions and fluids in the stomach. The secretion, urine, being acid, counter currents arise and are produced by the reaction of the acid of the urine upon the fluids and substances with which it comes into contact. In judging, therefore, of the effects upon the needle, we must take into consideration the *acting points* in the circuit; there may be at least three acting points in a circuit, viz. at the point of secretion, and at each of the electrodes. If the *direction* of the current consequent upon secretion coincide with those that occur at the electrodes, then an increased effect upon the needle is necessarily produced; but if these currents tend to go in opposite directions, then the result upon the needle will be merely the *differential* effect. Hence we should be led to very erroneous conclusions, judging merely from the effect upon the needle, either as to the *force* of the current, or its *origin*.

Sufficient evidence has been obtained to warrant the following deduction, viz. *that when the electrodes of a galvanometer are brought into contact, one with the urinary secretion and the other with the venous blood from the same part, an effect upon the needle occurs indicating the blood and the urine to be in opposite electric states.*

It may just be remarked, that slight effects may
be observed when the electrode is in contact with
the arterial blood instead of the venous blood, the
other being in contact with the urine. But no
effects are obtained when one electrode is inserted
into the vein and the other into the artery of the
kidney.

Whilst upon the subject of urinary secretion, I
may allude to a circumstance of some interest. At
the time the original experiments were performed,
it was frequently observed that the blood continued
to indicate its *positive* condition, long after the
secreting process could have been going on, which
led to the belief that the blood might have the
power of retaining its peculiar electrical state. Subsequent experiments have tended to confirm this
opinion, but it was never supposed that the secretions
could have the power of retaining their peculiar
electrical condition until lately. Reading over some
of the Memoirs published at the time of the celebrated controversy between GALVANI and VOLTA, I
was much gratified by accidentally finding a letter
written by VASSALI EANDI, at that time one of the
celebrated Professors at Turin, to M. DELAMETHRIE,
then Secretary to the Royal Academy of Paris, who
requested his opinion "*upon Galvanism and the origin
of Animal Electricity,*" and to which I have referred
in chap. i. The position these two individuals
held might be adduced as giving some weight to
their authority, Amongst other arguments that

Vassali Eandi brings forward in favour of the existence of Animal Electricity is the following : " J'ai prouvé ailleurs," says Vassali Eandi, " que les urines donnent une électricité négative, et j'ai fait voir plusieurs fois aux D. Gerri, Garotti et aux élèves de médecine et de chirurgie, que le sang tiré des veines donne dans mon appareil électrométrique (décrit dans le Vol. V^e de l'Académie des Sciences de Turin, Dec. 19, 1790) une électricité positive."

It need scarcely be stated, that the galvanometer was not then known, and that the effects observed by Vassali Eandi were those of *attraction* and *repulsion*. Although the results obtained by Vassali Eandi may be supposed to be due to other circumstances than those arising from Animal Electricity, such as evaporation or chemical action, nevertheless, as recorded facts, they are of some value, inasmuch as they tend to establish similar conclusions which have been arrived at by different modes of investigation, and entirely independent of each other.

Sect. IV. *On the Manifestation of Current Force during Mammary Secretion.*

In my original paper only one experiment was recorded, as shewing the results that were obtained in the Mammary gland; since then, several other opportunities have occurred in which similar results were observed.

If we insert the electrodes, one into a lactiferous vessel and the other into a vein from the same part, the electrode in contact with the vein is *positive* to the other 8° or 10°.

Here, in this instance, we get evidence of *the secreted product (the milk) and the venous blood being in opposite electric states.*

It will now be seen, from the foregoing experiments, that wherever secretion occurs, whether in the stomach and intestines, in the liver, in the kidneys, or in the mammary gland, it will be found that the act itself is not only accompanied with the manifestation of *current force*, but that the *venous* blood is also, in all these instances, in the same state, in a *positive* electric state. The next question that would naturally arise is the following : What is the state of the *arterial* blood ? Although it has been found that the arterial blood indicates a *positive* state when formed into a circuit with the secreted product, the other necessary element, viz. its *electronegative element*, the *cation*, for example, has not yet been obtained. Reasoning from analogy, it is in the lungs that a satisfactory explanation on this point must be sought for.

Physiologists may not perhaps be disposed to admit that the function of the lungs corresponds to that of a secretory organ; or that the process by which carbonic acid is eliminated from the blood corresponds to that by which the acid is eliminated from the stomach ; fortunately a decision on this point will not be necessary, and therefore need not detain us. One circumstance, however, is well known, viz. that *carbon*, in some form or other, is eliminated from the blood during its passage through the lungs; and it may so happen, that during the elimination of this carbon, its *separation* from the venous blood whilst traversing the lungs, that current electricity becomes manifested.

SECT. V. *On the Manifestation of Current Force during Respiration.*

When one electrode is brought into contact with the mucous membrane of the bronchial tubes, and the other inserted into the left ventricle of the heart, the latter electrode is *positive* to the former from 2° to 5°. When the electrode, instead of being inserted into the left ventricle was inserted into the right ventricle, it still indicated a *positive* state. Here then are indications of the arterial blood being *positive* to the mucous surface of the lungs: how far this state may be due to the *separation* of the *carbon* from the *venous* blood which traversed that organ may be a subject of dispute; the fact, however, is of some importance, as indicating the electric condition of the *arterial* blood.

In looking back upon the results that have now been obtained, some surprise may be felt at the circumstance, that all these experiments tend to indicate that, during life, the blood, whether *venous* or *arterial*, is in a *positive* electrical state or condition, and that this state or condition is partly produced and maintained by the various secretions that take place in the animal body. How far the fluidity of the blood, and the vitality of the blood, as it is called, are dependent upon this electric state or condition, are questions which must necessarily arise in our minds. The particles of the blood, also, must under these circumstances exist in a state of self-repulsion : and may not this fact, it may be asked, tend to explain some of the phenomena connected with the circulation of the blood in parts not dependent upon the *vis à tergo* action of the heart, and also those connected with the coagulation of the blood when taken from the living animal ? Before concluding this Section let me just observe, that a clue has now been obtained to the non-appearance of any effect upon the galvanometer when the two electrodes are inserted into an artery and a vein, a point previously established by the experiments of POUILLET and MULLER[k]. As the blood in the two vessels is in the same electric state, no effect could occur upon the needle ; thus proving the fact, well established by FARADAY, that in order to obtain CURRENT FORCE, the *circuit form* must be given to the arrangement, *i. e.* that *the electrodes must be brought into contact, or by means of some conducting mass, with the* ANION *and* CATION *originating the power*[l].

There are one or two points which must now be noticed. It may be supposed, *first*, that the effects

[k] *Loc. cit.*   [l] *Experimental Researches*, vol. ii. p. 51.

that have been obtained may arise from *thermo-electric actions*, since Becquerel[m] and Breschet have ascertained the existence of a difference in temperature between the *arterial* and *venous* blood by means of a galvanometer; *secondly*, that they may also arise from the actions that take place upon the surface of the platinum electrodes. We will now notice these objections.

First, as to *catalytic actions*, or *the combining power of platinum*. There are strong experimental reasons for believing, that when blood escapes from a wound, it enters into *combination* with the oxygen of the atmosphere; when a plate of platinum therefore is in contact with the blood, actions similar to those which occur in the gas-battery take place. We have a right to suppose that similar actions would occur at the other electrode, namely, that in contact with the bile; still, it might not necessarily follow that the latter would counteract the effects of the former. Judging then from the *direction* of the current, the effects may be fairly supposed to be due to the actions which occur between the atmosphere and the blood, or, in other words, to *catalytic actions*.

There can be no doubt that the effects observed are partly due to *catalytic actions;* and we may even go further and say, that they *must* be so as a necessary consequence. Just now it was necessary to suppose the blood to be *acid*—to contain an *anion*—to account for the effect; now we are obliged to suppose it to

[m] *Loc. cit.* tom. vii. p. 20.

contain a *cation*—to be alkaline—to account for the catalytic action, or if not in an alkaline state, still in such a state as to combine readily with the oxygen of the atmosphere. Now this latter state we are necessarily driven to entertain, when supposing, which no physiologist will deny, that the blood, during secretion, undergoes a change similar to that of *decomposition*[n].

Secondly, as to *Thermo-electric actions*. BEC-. QUEREL[o] and BRESCHET, as is well known, have shewn that different parts of a living animal are of different temperatures; but it must be borne in mind, that their experiments were intended to elucidate thermo-electric actions, and might not, therefore, be regarded as comparable with the present. Although it would be considered rather a stretch of the imagination to suppose that the effects can be referred to thermo-electric actions, since no effect was obtained when the electrodes were inserted into the vena porta and hepatic veins, as in former experiments, or even in the experiments of MULLER, between the corresponding arteries and veins, still, it is for physiologists to shew that the effects cannot be referred entirely to these actions. To assist in deciding this question, the following experiments were undertaken.

A porcelain jar, 2 inches and a half in diameter

---

[n] It is not necessary to point out in what manner, whether by parent-cells or secreting-cells.

[o] Traité de l'Electricité, tom. vii. p. 20.

E

and the same in depth, capable of holding about five ounces and a half of fluid, was used as the external cell; a portion of the small intestines of a rabbit, eapable of holding half an ounce of fluid, was suspended by threads, and formed the internal cell; the ends of the electrodes, to the extent of half an inch, were bent at a right angle, and placed in each cell, the other extremities being connected with the galvanometer and mercurial cups, as in the experiments on animals. Thus arranged, water at different temperatures was poured into each cell.

Experiment 1.

Temperature of atmosphere . . .	71°
Temperature of water in external cell	68
Temperature of water in internal cell	120

Slight vibrations of the needle. Every endeavour to obtain a greater effect failed. The temperature of the fluid in each cell was then ascertained by means of a delicate thermometer.

Temperatnre of internal cell . . .	105°
Temperature of external cell . . .	80

Experiment 2.

Temperature of external cell . . .	160°
Temperature of internal cell . . .	68

Vibrations as before; and it was then found that the

*Temperature of external cell was .	125°
Temperature of internal cell . . .	98

Experiment 3.

		o
Temperature of external cell	. . .	67
Temperature of internal cell	. . .	130

Vibrations; and

		o
Temperature of external cell	. . .	81
Temperature of internal cell	. . .	110

In whatever manner the experiments were varied, whether by using water at greater or less differences of temperature, similar results were obtained. The vibrations were sharp and quick at the commencement, but soon terminated; in no instance could a decided effect upon the needle be obtained by making and breaking contact, and the effects were not in any way similar to those observed in the animal body. There is one remark, however, which might be made in reference to these experiments, viz. if the two fluids could be kept at constant temperatures at the point of contact, more decided effects might be expected.

In the following experiments a resistance—a liquid conductor—was added to the circuit, to see if the current would be capable of traversing it. A glass tube, nearly half an inch in diameter and 3 inches in length, was bent thus, **U**, and contained water; one limb was connected with one of the mercurial cups by a piece of copper wire of the same thickness as those connected with the galvanometer, and 3 inches in length; the other limb of the tube was connected with another mercurial cup by a similar

piece of copper wire; each of these wires dipped, to
the extent of a quarter of an inch, into the water.
We thus had a resistance consisting of a column of
water, nearly half an inch in diameter, 2 inches and
a half in length, and 6 inches of copper wire. By
this arrangement the current could be made to
travel through the galvanometer, with or without the
resistance, at pleasure, by merely dipping the electrodes into one mercurial cup or the other, and
without any loss of time.

Experiment 1.—Rabbit. Pithed. Between renal
vein and bladder; *with* resistance, vibrations; *without*,
3° or 4°: *with*, vibrations; *without*, vibrations.

Between left bronchus and left ventricle; *with*
resistance, no effect; *without*, 4°: *with*, 2° or 3°;
without, no effect.

Between gall-bladder and blood from vena cava
inferior; *with* resistance, 5°; *without*, 8°: *with*, 5°;
without, 10°. The motion of the needle with the
resistance was slow and steady.

Experiment 2.—Rabbit. Pithed. Between right
bronchus and left ventricle; *without* resistance, 2° or
3°; *with*, vibrations: *without*, vibrations.

Between gall-bladder and blood from vena cava
inferior; *without*, 10°; *with*, 3° or 4°: *without*, 10°.

Between renal vein and bladder; no effect either
with or *without* the resistance.

I will not attempt to deduce any conclusions
from these experiments as to the *force* of the current,
but leave it for the physical philosopher to decide

whether the effects can be entirely due to thermo-electric actions. But there can be no doubt that a part of the effects may be referred to these circumstances, and partly also to *catalytic* actions, and they must therefore be taken into consideration when judging of the final result upon the needle.

The results recorded in the present chapter tend to establish the following conclusion, viz. *that the act of secretion in the living animal is accompanied with the manifestation of* CURRENT FORCE; and the phenomena with which this act of secretion appears to be the most intimately related are those that occur in the voltaic circle. A difficulty may arise to some minds in perceiving this relation, from the circumstance that in the ordinary voltaic circle metals are employed. If we bear in mind that the metals, although one of them is usually acted upon, serve principally as *conductors*, and that they are not *essential* for the developement of the power, this difficulty will be easily removed. Now as the manifestation of *current force* during the actions which occur in the voltaic circle are considered as evidence of *polar* action, there can be no reason why it should not be so considered in regard to organic action, viz. during secretion; but before we arrive at this conclusion, let us compare the phenomena of secretion with another class of facts, viz. with those of *osmose*.

Professor GRAHAM has communicated a very valuable Paper to the Royal Society, entitled, *On Osmotic*

Force, which has since appeared in their *Trans-actions*[p]. In this Paper Professor GRAHAM has shewn that *osmose* is dependent upon *chemical action*, and not, as it has been generally supposed, upon *capillary attraction*. Space will not allow me to enter upon the facts brought forward in support of this opinion; and I must therefore refer my readers to the Paper itself, which cannot be too strongly recommended.

The conditions under which an *osmotic* experiment is conducted, viz. the necessity of having two fluids, one on each side of the septum, render it extremely difficult to ascertain by means of the galvanometer the exact mode of action which arises during *osmose*, so as to compare it with that which takes place in the animal body during *secretion*, in consequence of the reaction of the two fluids upon each other pro-ducing their own peculiar effects on the galvanometer; and the changes upon which *osmose* depends take place, according to Professor GRAHAM, *within* the substance of the porous diaphragm, where we cannot apply the electrodes of the galvanometer.

The fact of *osmose* depending upon chemical action, shews however that the act itself must not be considered as a mere transudation, a mere physical separation, but that it depends upon other important conditions; and if upon chemical action, they are consequently polar in their nature. If this con-clusion be arrived at in regard to osmotic phenomena, we may with equal propriety consider the phenomena

p *Phil. Trans.* 1854.

connected with secretion to be at least something more than a mere physical transudation; and as reasons exist for shewing that osmotic phenomena are polar in their nature, why may we not also consider the action connected with secretion, and where we can obtain such direct evidence of polar action as manifested by the galvanometer, to be polar in its nature?

Respecting the chemical character of osmose, and its bearings upon physiology, Professor GRAHAM adds: "It may appear to some that the chemical character which has been assigned to osmose takes away from the physiological interest of the subject, in so far as the decomposition of the membrane may appear to be incompatible with vital conditions, and osmotic movement confined therefore to dead matter. But such apprehensions are, it is believed, groundless, or at all events premature. All parts of living structures are allowed to be in a state of incessant change—of decomposition and renewal. The decomposition occurring in a living membrane, while effecting osmotic propulsion, may possibly be of a reparable kind. In other respects, chemical osmose appears to be an agency particularly well adapted to take part in the animal economy."

CHAP. V.

FROM the results that have been detailed in the last chapter, I was naturally led to suppose, that during *lacteal absorption* some evidence might be obtained of the manifestation of *current force;* since an identity between this process and secretion has been long recognised by physiologists. MULLER[b] says, " *absorption (lacteal)* seems to depend on an attraction, the nature of which is at present unknown, but of which the very counterpart, as it were, takes place in secretion; the fluids altered by the secreting action being repelled towards the free side only of the secreting membranes, and then pressed onwards by the successive portions of fluid secreted."

---

[a] My original Papers, on the Developement of Current Force during Lacteal Absorption, Nutrition, in the Muscular and the Nervous Tissues, and in Plants, were communicated and read before the Royal Society in 1852. It being decided that they should not be published in the Transactions, they were ordered to be placed in the Archives of the Society. Upon wishing to refer to these three Papers two years afterwards, they were *missing* from the Archives of the Society, and were ultimately found to be in the hands of the referee !!!

[b] Elements of Physiology, translated by Baly, 2nd edition, vol. i. p. 301.

The experiments were performed upon cats, guinea-pigs, and rabbits; young cats will be found to be the best subjects. After the animal had been kept without food for some hours, a large quantity was then given to it, and the experiment performed at different periods, from a quarter of an hour to an hour, or longer, after the food had been taken. According to the time that elapsed between the meal and the experiment and the previous empty state of the stomach, so could the lacteals be traced out in various parts of the intestines; and it was interesting to observe the marked contrast between the empty portions of the gut, and that portion where digestion was going on; the difference in the vascularity of the parts being so very great. One electrode being inserted into the intestine, and the other in contact with the chyle flowing from a lacteal coming from the same part, the latter was *positive* 8° or 10°. The effects were just the same as if the electrode had been in contact with the blood from a vein from the same part as described in the last chapter.

Now it may be said, and with some justice, that these results are evidently due to *secretion;* that they are the identical experiments that have been related in the last chapter, the only difference being, that the *force* is conducted by the mesentery. In reply to this it may be stated, that *those* effects, instead of being referred to *secretion*, may be considered as due to *lacteal absorption.*

To decide this point, let us just observe we have two facts before us. *First,* the venous blood indicates a *positive* condition; and, *secondly,* the chyle also indicates a similar state. Now if any difference exists between these two fluids, we shall detect it by inserting the electrodes, one into the vein, and the other into a lacteal; but upon doing so no effect[c] occurs upon the needle. This experiment is the very counterpart to those of POUILLET and MULLER, when they inserted the electrodes into the vein and artery of a living animal without observing any action upon the galvanometer, and to which I have already alluded. This negative result would only prove that the chyle and the venous blood are in the same electric states; and all that can be said in regard to the experiments we are now considering is this, that the effect, the *positive* state as indicated by the needle, may be due partly to the *positive* state of the chyle, and partly to that of the *venous blood* which is conducted by the mesentery.

From what has been already remarked in regard

---

[c] It must not be supposed that *no effect* is *always* obtained under these circumstances. It is difficult and almost impossible, in these physiological experiments, not to obtain some slight effect upon the needle; but any one, after a little experience, would even be enabled to ascertain the marked difference between a decided and definite action upon the galvanometer, due to a definite and permanent cause, and that which might arise from some slight and variable circumstance, such as a difference in the state of the parts as regards moisture, and as to the contents of the intestines, &c.

to the acids in the stomach, it may be readily supposed that in this viscus the same effects would be observed as in the former experiments; viz. that the current dependent upon the acid secretions would overcome and mask any effect that might arise from lacteal absorption.

Sufficient evidence, I think, has been adduced to prove the following conclusion, viz. *that during lacteal absorption the absorbed product* (the chyle) *and the substances from which it is absorbed* (the food) *are in opposite electric states.* The conditions necessary for the manifestation of *current force* are complete; we can trace out the *anion* and *cation* of the circuit, the former in the food, the latter in the *chyle;* we have no reason to believe that the chyle is *acid* in order to account for the effects, for all chemical analyses tend to shew its alkaline condition; and I need not repeat the arguments employed in the last chapter in regard to the supposed acid condition of the blood, for the purpose of accounting for the effects as they are applicable on the present occasion. I would not go so far as to say that acids are never to be discovered in the chyle or in the blood; for it is possible that in some morbid states of the system this circumstance might occasionally happen.

I am not now attempting any explanation as to the mode by which *lacteal absorption* takes place or is brought about, whether it is to be considered as a mere case of *imbibition*, or as a case of *osmose.* I am only shewing that during this act *current force* is manifested, and that consequently it is a *polar* phenomenon. Perhaps it would be as well for the present to consider it as an *organic* act, and it may be looked upon as a case of *organic endosmose*, whilst secretion, on the other hand, may be considered as a case of *organic exosmose*.

CHAP. VI.

I HAVE now arrived at a point of my subject which has unfortunately been a source of some dispute and bitter controversy; into this, however, it will not be necessary for me to enter: and as my chief object will be to point out the *origin* or *cause* of the principal effects observed, rather than to give an elaborate and detailed account of all the facts that have been obtained, I shall not go so fully into the subject as might be expected, but endeavour to be as concise as possible, not omitting any thing of importance.

The question, whether Nutrition is accompanied with the manifestation of *Current Force*, will be considered under two heads. *First*, in regard to the Muscular Tissue; and *secondly*, in regard to the Nervous Tissue.

SECT. I. *On the Manifestation of Current Force during Nutrition in the Muscular Tissue.*

I have already alluded to the circumstance in chap. i. that so far back as 1827, NOBILI, in some experiments undertaken for the purpose of ascertaining the comparative sensibility of the galvanometer and the galvanoscopic frog, found, that when the frog's limb was so arranged that the feet were placed in one vessel containing water, and the lumbar nerves in another vessel, then upon completing the circuit with the galvanometer an effect was produced upon the needle indicating a current to pass from the feet to the upper part of the limb; the electrode in the vessel containing the nerves being *positive* to the other. This he called the *proper current* of the frog, a term by which it is designated at the present day, and he considered that the effects were thermo-electric in their nature.

MATTEUCCI subsequently pursued the subject, and amongst other most important results he ascertained, that when a muscle was divided and one electrode placed in contact with the divided surface, and the other in contact with the external surface, an effect was produced upon the needle. He further shewed, that this effect depended upon the vital condition of the muscle, upon the state of its nutrition.

After the publication of MATTEUCCI's earlier experiments, DU BOIS REYMOND took up the subject, and from a series of ingenious experiments deduced what he designated as the law in regard to the *muscular current*, viz. that any point of the *longitudinal* section or the surface of the muscle was *positive* to any portion of the *transverse* section, whether natural or artificial, of the same muscle. Finding, however,

that this law does not hold good in its entire gene-
rality, that the current was not *always* obtained when
the so-called *natural transverse* section (viz. the
tendon) and the *longitudinal* section were formed
into a circuit, he was led to suppose that a layer,
which he denominated the *para-electronomic* layer,
exists, and which might fully account for the occa-
sional non-appearance of the current. For my own
part, I have never been able to satisfy myself of the
existence of this layer or its necessity; but there can
he no doubt about the fact, that when the *divided*
surface or *artificial transverse* section and the *longitudinal* section or surface are formed into a circuit,
that then the so-called *muscular current* is obtained.

Previous to any knowledge of Du Bois Reymond's
researches, I was led, from the results that I had
already obtained in regard to *secretion*, to helieve, that
if the tissue, the muscular, was formed into a circuit
with the venous blood, I might perhaps ohtain
some evidence of ·a current, as I had already done
in regard to secretion. I was further strengthened
in this supposition, by finding that Matteucci
explained the results he had obtained by repre-
senting the muscular fibre as similar to a zinc plate
in a voltaic circle. He says, " We must never forget
the analogy between the muscular electro-motor
element and the voltanian element: the zinc is
represented by the discs of the muscular fibre, the
acid liquid hy the blood, the platinum by the sarcolemma. Whatever he the conducting body with

which the zinc is made to communicate with the platinum, the current is always in the same direction...... The chemical actions of nutrition evolve electricity[d]." It has always been a source of great surprise to me, that MATTEUCCI has never attempted, as far as I am aware, to ascertain the correctness of this opinion or the result of the experiment, viz. that of forming a circuit between the muscular fibre and the blood, as it was the only evidence required to prove his conclusion.

I now performed the experiment in the following manner. Laying bare the muscles of the animal at the upper and inner part of the thigh, one electrode was inserted into the substance of the muscle, or, which is better still, placed in contact with the divided transverse section of the muscle, and the other electrode was inserted into the blood flowing from a neighbouring vein; a slight effect occurred upon the needle indicating the blood to be *positive*. If the electrode be placed on the external surface of the muscle instead of its divided surface, the effects are doubtful. If both electrodes were in contact, one with the divided surface of the muscle and the other with its external surface, then the latter electrode was *positive* to the former; the result being the so-called *muscular current*. That the effects should be null or difficult to be obtained when the external surface of the muscle and the venous blood are formed into a circuit, need not be surprising, since

[d] Philosophical Transactions, 1845, p. 301.

the *positive* state of the surface of the muscle prevents that of the blood from becoming manifested, these two parts, in fact, presenting the same electric condition.

I need not relate other experiments proving the conclusion, that we can obtain evidence of current force being manifested when the muscular tissue and the venous blood are formed into a circuit. The effects, it is true, are but small, but nevertheless definite; and the question now arises, can they be referred to nutrition?

Although it may perhaps be difficult at first sight to see that the effects can be referred to nutrition, nevertheless, when we bear in mind the strong resemblance that exists between secretion and nutrition, this apparent difficulty will be removed. Nutrition differs from secretion in this; in nutrition the secreted product (the tissue) remains behind, and is not carried off as in ordinary secretion. Not only does the tissue (the secreted product) remain behind, but it also retains its peculiar electric *negative* state, and may be considered as the *anion* of the circuit, whilst the *cation* exists and passes on in the blood. The electric state is not only produced by nutrition, but maintained by it. We have no right to suppose that the blood is *acid* in order to account for the effects; and it would appear, according to LIEBIG's researches, that *lactic acid* is developed in the muscular tissue.

I am not going so far as to suppose that nutrition,

F

as it takes place in the muscular tissue, is so rapid a change as may occur in some of the secreting organs, that the tissue is deposited and immediately taken up again, or constantly undergoing changes; I am far from supposing this, and do not pretend to be able to trace out the exact changes that take place, nor the time that is required to effect these changes. We have, however, under any point of view, to consider how this electric condition of the tissue is brought about, and what other circumstance, I may ask, is more likely than nutrition? We have a *vera causa* at work, and such a one as is likely to produce the very result observed. I may also refer to the results obtained by MATTEUCCI, in which he has shewn the dependence of the *current (muscular)* upon the state of the tissue in regard to its nutrition, that whatever influences the latter affects the former. At the same time, I am perfectly aware that objections may be raised to his mode of conducting the experiments, viz. by forming piles of animal substances, but I cannot help thinking that these objections may be carried too far.

It may be asked, and with some justice, how can we account for the external surface of the muscle being *positive* to the divided transverse surface? we can understand, it may be said, the *cation* existing in the blood producing its *positive* state, but where is the *cation* in the external surface?

I shall allude to this question in a future chapter, when it will again come under consideration; and as

I am now going to state some conclusions, the proof of which have not yet been presented to my readers, I must refer to this chapter for the requisite evidence. My chief object now is to shew, that *nutrition* is accompanied with the manifestation of current force; and the fact that the tissue is in an electric state, may be considered as a corollary to this conclusion. The *current* (*muscular*) therefore may be considered as existing under two points of view : *first*, when the tissue and the blood are formed into a circuit, the *current*, then manifested, is due to the *act* of nutrition, it continues *during* nutrition, whilst the *changes* are going on, just as the current is manifested in the voltaic circle during the time that the chemical changes are taking place : *secondly*, when the *external* and *divided* surfaces are formed into a circuit, then the effects resemble those of a charged Leyden jar rather than those of a voltaic circle. The tissue itself being *negative*, and what is more was deposited as such *during* nutrition, the external surface, the sarcolemma, becoming *positive* by induction.

Before passing to the consideration of the nervous tissue, I think sufficient evidence has been adduced to warrant the following conclusion, viz. that *during nutrition in the muscular tissue, the muscular fibre and the venous blood are in opposite electric states.*

SECT. II. *On the Manifestation of Current Force during Nutrition in the Nervous Tissue.*

DU BOIS REYMOND was the first to shew that the nervous fibre, like the muscular fibre, presented an electrical condition; the external or longitudinal surface being *positive* to the divided surface, producing the so-called *nerve current*, when the two surfaces are formed into a circuit with the galvanometer.

The results that I had obtained in regard to the muscular tissue led me to make inquiries respecting the nervous tissue; and for this purpose, after having exposed the brain, I inserted one of the electrodes into the substance of the brain, and the other into the internal jugular vein; the effects upon the needle were very decided and manifest. The needle advanced from 10° to 20° or more, the electrode in contact with the blood being *positive* to the other in the brain.

MATTEUCCI[e] relates some experiments performed by MM. PACINOTTI and PUCCINOTTI, and by himself, in which the electrodes were inserted one into the brain, the other into the muscles. He says, "La déviation obtenu dans la première immersion a été dirigé du cerveau aux muscles dans l'animal. L'intensité du courant est très-variable : j'ai obtenu quelquefois 80° et même davantage, et quelquefois 10° à 15°, et toujours dans la première immersion."

[e] Traité des Phénomènes Electro-Physiologiques, p. 121.

The object of these experiments was, I believe, for the purpose of ascertaining whether any electric current traversed the nerves. I shall have occasion to allude to them again in a subsequent chapter. Why the effects should be so decided in this instance, when compared to those that are observed in muscles, may arise from the greater mass of the brain, and to the circumstance of the blood coming directly from it, and thus the conditions may be better adapted to shew the effect, than when the muscle and vein were formed into a circuit in the thigh. If the experiment be performed with the sciatic nerve, placing one electrode in contact with the divided surface, or the external surface, and the other in contact with the blood coming from the vein of the nerve, then the effects are similar to those observed with the muscles. Repeating Du Bois Reymond's experiment, viz. forming the divided and external surfaces into a circuit, the same result was obtained, the external surface was *positive* to the other; but the effects were not so great as when the same parts of a muscle were formed into a circuit; nevertheless the effects were decided and definite.

I need not repeat the arguments that were used respecting the blood being *acid*, in order to account for the effects; and with regard to the external surface of the nerve being *positive*, to the divided or transverse section; this may be considered as due to the same circumstance as was observed with the muscular fibre, viz. the *positive* state being *induced* by the *negative* condition of the nerve fibre. I may just remark, that the very fact of the muscular and nervous tissue presenting the same results, would tend also to confirm the opinion, that they are evidently due to one and the same cause, viz. *nutrition;* and I think sufficient evidence has been adduced to shew, that *the nervous tissue and the blood (venous) are in opposite electric states.*

Before quitting this part of my subject, I cannot too strongly urge the necessity of constantly bearing in mind the importance of attending to the circumstances under which the current is manifested in these two tissues, the muscular and the nervous; and as I have just before alluded to this point when speaking of the *muscular current*, I must refer to the remarks there made, as they will be found applicable to our present case, to that of the *nerve current.*

CHAP. VII.

" Should any physiologist," says Muller, " be so
fortunate as to prove beyond doubt the electric
property of the blood, I could only congratulate
Science on the great advance which it would thus
have made[a]." From these observations of Muller
it would appear, that up to his time no satisfactory
conclusions had been arrived at in regard to this
question.

From the experiments related in the previous
chapters, besides the conclusions then drawn in
regard to the manifestation of *current force* during
secretion and nutrition, we may also deduce the
following conclusions as corollaries.

First, *That the blood, both venous and arterial, while
circulating in the animal body, is in a positive electrical
state;* and,

Secondly, *That this electrical state is produced and
maintained by the secretory, respiratory, and nutritive
actions.*

Now these conclusions which have been arrived
at as corollaries to other conclusions go far to prove

[a] Muller's Physiology, translated by Baly, 2nd edition, vol. i.
p. 148.

the experimental results obtained by VASSALI EANDI,
and to which I have already referred[b]. To him,
therefore, must the credit be given, as having been
the first to establish the important fact, the electric
condition of the blood by experimental evidence.
That his experiments should have been overlooked,
or even passed by as undeserving of attention, is
somewhat remarkable ; and I can only refer it to
that injudicious scepticism, too frequently expressed
where a knowledge of the subject is wanting, and
sometimes from other motives, which is as great a
bane to the progress of science, as the opposite evil
a too easy credulity.

Although objections may be raised to the experimental results obtained by BELLINGERI as recorded
by MULLER, he nevertheless appears to have arrived
at the conclusion, that arterial and venous blood
presented no difference in regard to their electrical
properties ; and that the electrical property of the
blood is preserved long after its abstraction from
the vessels.

There can be no doubt that most important
phenomena are connected and associated with this
electric condition of the blood; that a great many
of its properties, which have been referred, in consequence of being inexplicable, to a *vital* condition,
belong to it. Its particles must be in a state of self-
repulsion, and may therefore assist in the circulation
of the blood. Some of the peculiarities connected

[b] Chap. i.

with the coagulation of the blood are no doubt also dependent upon it; but as these are questions which require special investigations for their solution, I shall merely allude to them, and observe, that the subject presents to any inquirer an ample field for future discoveries, important to the pathologist as well as to the physiologist.

CHAP. VIII.

In chap. vi. besides the conclusion that current force was manifested *during* nutrition in the muscular and the nervous tissues, and that the tissues, the muscular and the nervous, presented one electric state, the *negative*, whilst the venous blood presented the other condition, the *positive;* it was also shewn, from the experiments of MATTEUCCI and Du Bois REYMOND, and confirmed by some of my own, that similar effects were obtained when the external and divided surfaces, either of the muscle or of the nerve, were formed into a circuit producing the so-called *muscular* and *nerve* currents, the external surface in both instances being *positive* to the divided or transverse section. The question naturally arises as to the *origin* of the current; we can understand the *negative* state being produced *during* nutrition, the tissue in fact being deposited as the *anion*, whilst the *cation* is passing away in the blood; but, it may be asked, where is the *cation* on the external surface? MATTEUCCI[a], in some of his recent

---

[a] Bibliothèque Universelle de Genève, &c. 1856.

researches, appears to entertain some difficulty in considering what really is the true electro-motor element of the circuit under these circumstances. I have already suggested, that the muscular tissue and the nervous tissue may be considered in the same light as a charged Leyden jar; that the *positive* condition of their external surfaces may be caused by *induction*, the tissue itself being *negative*, and deposited as such during nutrition. Before we can come to this conclusion, however, it will be necessary to investigate the subject more fully.

For the convenience of discussion, I may just sum up, in a few propositions, the results which appear to be well established by the researches of MATTEUCCI and DU BOIS REYMOND, and such as I have been enabled to confirm by my own experiments, which it will not be necessary to particularize.

1. *Any point of the surface of a muscle is positive in relation to any point of the divided or transverse section of the same muscle.*

2. *Any point of the surface of a nerve is positive in relation to any point of the divided or transverse section of the same nerve.*

These two propositions express the law as deduced by DU BOIS REYMOND in regard to the *muscular* and the *nerve currents*.

DU BOIS REYMOND has, however, applied the terms *natural* and *artificial* to the different sections. There may be no objection to the employment of these terms; but unfortunately the so-called natural transverse section may act as the surface or longitudinal section in regard to the artificial section or the divided surface; and I do not think that the existence of the *para-electronomic* layer can be sufficiently made out, to account for the difference of effect which may thus occur to justify us in employing these terms; nevertheless, I perfectly agree with Du Bois Reymond, in considering that the *side* or *external surface*, or perhaps the sarcolemma of the muscle, and the *base* or *ends* of the muscular fibre, are the important points to be in contact with the electrodes for the production of *muscular* current, and the similar parts in the nerve for the production of the *nerve* current.

3. The *intensity* of the *muscular* and *nerve* current depends upon the vital conditions of the tissues; the *current* does not subside immediately. after the death of the animal, and evidently bears some relation to the state of its nutrition.

4. According to Matteucci[b], the *intensity* of the muscular current varies as the *length* of the muscle in the circuit, and not according to the extent of its *transverse* section. This proposition is, however, rather difficult to prove; for, if two muscles of the same length, but of different thicknesses, be formed into a circuit, the thicker muscle will be found to give the more intense current, (and the same will be found in regard to the nerve;) and I cannot help thinking, that Du Bois Reymond is correct in con-

[b] *Loc. cit.*

cluding that the electro-motive force of muscles increases both with their length and thickness.

5. If the two ends—the divided or artificial transverse sections—of a muscle or of a nerve be placed between the electrodes of a galvanometer, no effect occurs upon the needle; but if one of the electrodes be placed upon the surface or longitudinal section, the other remaining in contact with the transverse section, the current is produced according to the law expressed in Proposition I.

6. If the electrodes be placed upon two symmetrical portions of the surface of the muscle or of the nerve, no effect is produced upon the needle.

The following experiments were now performed: A muscle or nerve was divided into three or more portions, and then so arranged that the internal should form the outer portions, the continuity of the nerve being maintained. When the electrodes were placed at the two extremities, or upon symmetrical portions of the nerve or muscle, no effect, or, if any, but very slight indications were occasionally produced upon the needle; if the external surface and the divided surface were formed into a circuit, the current was obtained; and also when the portions were so arranged, that the external surface of one portion was placed in contact with the transverse section of another, then the current was obtained, as in MATTEUCCI's experiment, when forming a pile of muscular elements. These experiments, which I

consider of some importance, were repeated several
times upon the muscles and nerves of frogs, guinea-
pigs, and rabbits; care being taken to have the
portions of nerve and muscle, as far as possible, of
the same size, and from the same nerve or muscle.
They go far to shew, that the muscular or nerve fibre
does not indicate any manifestation of *polarity* in
regard to its *length;* there is no fact or evidence of
any kind to shew either that the nerve fibre or the
muscular fibre presents a condition at one extremity
opposed to that of the other extremity; there is no
antithetical or *dual* condition made manifest indicative of *polarity;* whatever indication of *polarity*
exists, shews it to be manifested in the *transverse*
direction, or rather that the *muscular fibre* or *nerve
fibre* presents one condition, and perhaps the *sarcolemma* or the *neurilemma* the other; and we are now
brought to the consideration of our original question,
Can this state be dependent *entirely* upon the
changes which occur *during* nutrition? And where
is the *cation?* Or can we suppose the sarcolemma
or the neurilemma to act as an *acid!* or might not
the effect (the current) be due to the *electrical*
condition of the muscular or nerve fibre, its *negatively*
electric state being maintained so long as the vital
conditions of the tissue continue?

Under the supposition that the effects might be
due to the *heterogeneity* of the parts, the sarcolemma
or the neurilemma acting as an *acid,* I now repeated
the experiment, employing the following dilute

solutions:—Five drops of strong sulphuric acid to one ounce of distilled water, formed the *acid* solution; one drachm of liq. potassæ (Phar. Londin.) to one ounce of distilled water, formed the *alkaline* solution; and one drachm of common *salt* to one ounce of distilled water, formed the *neutral* solution.

Whenever one electrode was moistened with the *acid* solution, this was *positive* to the other on whatever parts of the muscle or nerve they were placed. When *both* electrodes were moistened with the solution, the indications were doubtful.

When the *alkaline* solution was employed in the same manner, the electrode in contact with the external surface of the muscle or nerve was *positive* to the other; it very rarely happened that the electrode in contact with the tranverse section indicated a *positive* condition, when the other electrode alone was moistened with the solution.

With the solution of *salt*, similar effects were observed as with the *alkaline* solution; and from these facts we cannot but infer that these two solutions, the *alkaline* and the *salt*, acted merely as a conducting liquid, whilst the *acid* solution acted chemically upon the animal substances. It cannot be supposed, or at any rate be believed, that the sarcolemma or neurilemma acted, in these instances, as an *acid* substance, merely for the purpose of explaining or accounting for the existence and *direction* of the current, since the *alkaline* solution would have destroyed its *acid* reaction.

When the muscles and nerves had been placed in strong concentrated solutions, so as to act chemically upon the tissues, and different portions of them were then formed into circuits, the effects upon the needle were sometimes very powerful, sometimes null; the results, however, could never be predicated, and were evidently due to the chemical actions set up.

If the muscle or nerve was placed in hot water at the temperature of 180°, or in cold water at the temperature of 32°, so as to freeze it, the muscle or nerve current was seldom obtained.

If the muscle or nerve was squeezed up into a mass, so as to destroy its structure by mechanical means, no effect at all analogous to that of the muscular or nerve current was obtained; effects upon the needle were occasionally produced, but presenting quite a different character.

These results only tend to prove the conclusions already deduced by MATTEUCCI and DU BOIS REYMOND, of the dependence of the *muscular* and *nerve* current upon the normal or *vital* condition of these two tissues, and go far to shew, that these currents cannot be *entirely* due to the *heterogeneity* of the parts in the circuit, but that whatever destroys the normal or healthy state of the tissues, destroys also the conditions upon which the muscular or nerve current depends.

If the existence of the muscular and nerve currents be dependent *entirely* upon the changes which occur during the *act* of nutrition, it is reasonable to suppose that the removal of the blood from the limb might perhaps prevent them from being manifested. I was now led to the following experiments:

The animals—guinea-pigs, rabbits, and frogs—were bled to death, and the limbs and muscles then emptied, as far as possible, of the blood, by squeezing the limbs; under these circumstances the muscular and nerve currents were obtained. The amount of deflection was not perhaps so great as it would have been had the blood not been removed; nevertheless, the current was manifested. I endeavoured to remove the blood by injecting water (at the temperature of 90° in the guinea-pigs and rabbits, and at the temperature of the atmosphere for the frogs) into the blood-vessels; and although the water escaped by the veins, still the current was obtained, amounting from 2° to 5° in the muscles, and from 1° to 2° in the nerves. The muscles in these latter experiments became turgid, but they were not so pale as might have been expected; and I very much doubt whether the blood was entirely removed from the limb or the muscular and the nerve tissue. These latter experiments, as far as they go, would tend to shew the importance and dependence of the *current upon* nutrition; and although, in absence of further evidence, it would be impossible for us at present to state, or even conjecture, the period at which the act of nutrition terminates, there nevertheless appears to be a residual effect in these results which cannot be fully accounted for, under the supposition that the *current* is due *entirely* to the changes which occur *during* nutrition ; and I believe that we shall be justified in coming to the following conclusions, and in considering that the so-called muscular and nerve currents may depend upon three circumstances : *first*, upon the changes which occur *during* nutrition ; *secondly*, upon the *heterogeneity* of the parts, (including under this term the action of the platinum electrodes, viz. the *catalytic* action or the *combining* power of platinum upon the moist animal substances in contact with its surface ;) and, *thirdly*, upon an *electrical* state or condition of the muscular or nerve fibre itself.

In thus considering the muscular or nerve fibre as being in a peculiar state or condition which may be termed *polarized*[c], an objection might be urged to the employment of this term, inasmuch as the term *polarity* embraces the idea of *duality*—an *antithetical* action, which, as we have seen, does not exist in

---

[c] Dr. Todd, I believe, was the first to describe the true character of the phenomena I am now speaking of, and to point out their dependence upon nutrition, and also to consider them as being *polar* in their nature. See the article on the Physiology of the Nervous System in the " Cyclopædia of Anatomy and Physiology." Du Bois Reymond speaks of the tissue as being in an electro-tonic state. I have avoided using this term *electro-tonic*, as the facts appear to me to resemble those of a charged Leyden jar, rather than that of a wire conducting a current of electricity.

regard to the fibre itself. It is therefore necessary to point out with what class of phenomena the facts appear to be the most nearly allied.

Nutrition, I believe, may be referred to the same class of actions as *secretion*, the tissue, muscular or nerve tissue, being deposited as a secreted product. These actions, *secretion*, I have already considered as identical with those which occur in the *decomposing* cell of a voltaic circle, and to be POLAR in their nature. The question, however, whether the *secreted* product and the *blood* can, when separated, maintain their peculiar electrical states I have not yet examined; but in regard to the *blood*, I believe that that fluid *may*, from what I have occasionally observed in my former experiments, and confirmed as they are by the results obtained by VASSALI EANDI. At any rate, the improbability of the elements of compounds undergoing decomposition in a decomposing cell of a voltaic circle, retaining their peculiar electrical states, would be no positive argument against this supposition; the state and condition of the liquids in animals, in regard to their fluidity, and the conditions under which the changes occur, present great differences compared to those that take place in ordinary voltaic decompositions; consequently, it may be reasonably supposed, that under the circumstances in which the changes occur in the animal body, the electrical states of the solids and fluids are not immediately lost. From these facts, I feel no difficulty or hesitation in regard to the *origin* of the muscular or of the nerve current in the experiments we have been lately considering, viz. when the two surfaces of a muscle or of a nerve, the *transverse* section and the *external* surface, are formed into a circuit. *During* nutrition, the tissue is deposited as an *anion*, the *cation* passing on in the blood; the *negative* electric condition of the tissue is maintained by nutrition, and continues in that state; it may, perhaps, be looked upon as being in a state of tension, *polarized;* and when the muscular or nerve fibre, and any other part—for instance, the sarcolemma or the neurilemma—are formed into a circuit, an effect is produced upon the needle, indicating the existence of a *current.* Under these conditions, the results become analogous or are identical with those that are observed with a charged Leyden jar, when the inner and outer surfaces are formed into a circuit; the tissue being *negatively* electrized may, perhaps, render the sarcolemma or neurilemma *positive* by *induction*[d].

In speaking, therefore, of the fibre as being in a *polarized* condition, it would lead to erroneous views if we supposed that one portion, taking its longest diameter, was *positive* or *negative* to the other,

[d] This *positive* condition of the external surface will go far to explain a circumstance which at one time rather perplexed me. When the *external* surface of a muscle, and the *venous* blood flowing from it, were formed into a circuit, the effect upon the needle was but slight; the external surface of the muscle and the blood being *both* positive, will fully account for the feebleness of the current then manifested.

as a wire may be supposed to be when traversed by
an electric current; the fibre would appear rather
to represent an excited glass rod, as far as its mode
of action may be considered, rendering other parts
electric by *induction;* but whether the *force* exists
of that *intensity* to produce attraction or repulsion,
would appear doubtful after the experiments of
VASSALI EANDI[e]. It does not necessarily follow
that, because the *current* is produced, attraction and
repulsion should also be obtained; the *intensity*
requisite in the one case may be absent in the other;
and I cannot do better than refer to the important
Paper, by FARADAY, on the Gymnotus[f], for the purpose of pointing out the great difference manifested
in the two instances in regard to animal electricity
and machine electricity.   But the conditions under

[e] The following experiments of my own may, perhaps, be of
some interest. The limbs of frogs were suspended by means
of a silk thread to a glass beam, which was also suspended
horizontally by a silk thread attached to its centre, and fastened
above to another glass rod. Upon holding a silk thread near to
the limb, no attraction or repulsion of the thread was observed.
Exciting a glass rod, and presenting it to the limb, the limb
was powerfully attracted; it was attracted also upon presenting
an excited stick of sealing-wax to it. But I never obtained any
effect, when one limb was presented to the other, either of
repulsion or attraction; it would appear as if one of the substances must be in an *excited* state to produce these phenomena.
I do not consider these experiments of much value, and not at
all adapted to eliminate any very correct result; but they tend
to indicate the *negative* character of the tissue.

[f] Experimental Researches, vol. ii. p. 1.

which the muscular or the nerve current may occur
will be of the utmost importance to bear in mind:
whether as the *act* of nutrition, or *during* nutrition;
or whether as the *result* of nutrition, *i. e.* from the
polarized condition of the fibre itself, since they may
be referred to two distinct classes of actions.   Their
primary dependence, however, *upon* nutrition is a
circumstance of some importance; and it would
appear that this *polarized* condition of the tissue
is intimately connected with the so-called *vital* condition of the parts; the irritability of the muscular
fibre and the sensibility of the nervous tissue being
perhaps dependent upon it.

As the constituents of a muscle or its particles
must be in a state of self-repulsion, the electrized
or polarized condition not being limited to the
surface, as in a metallic conductor, but the whole
substance of the muscle throughout being equally
and bodily polarized, the question arises, Might not
muscular contraction be the necessary result of the
attraction between the particles, the muscular substance being, as it were, *depolarized* by *nervous* agency;
and the *force* by which the state of tension in the
muscle was maintained being evolved and set free
or made manifest in some other *form* or mode of
action, according to circumstances?   In a future
chapter we shall see that some evidence has been
obtained, indicating that, during *extraordinary* muscular exertion, some force *is* evolved, as in the fish;
but during *ordinary* muscular exertion, the *force* may

not become *free*, but be exerted in some other manner, as *heat* [s].

Before concluding, although I have alluded to the subject in chap. i., it may be remarked, that we are now brought to the consideration of the explanation of GALVANI's celebrated experiment, viz. the *contraction* that ensues when the nerve is brought into contact with the external surface of the muscle. GALVANI compared the muscles to a charged Leyden jar, the two surfaces, the external and internal, being in opposite electric states; and he considered that when the circuit was completed, the *contraction* was a necessary consequence of the passage of electricity from one surface to the other by means of the nerve. No one can deny that GALVANI was so far correct; he erred in considering the electricity as secreted in the brain, and transmitted by the nerves to different parts of the body, the muscles serving as mere reservoirs of the electricity; overlooking the fact, pardonable at his time, that electricity might be developed in other parts as well as in the brain. VOLTA erred in denying the *origin* of the power in the animal body, but was correct in pointing out that similar effects could be obtained by other means

[s] Dr. Radcliffe has published some views in regard to muscular contraction, in which he considers the muscle as being the seat of "*polar action*," and *contraction* as the result of molecular attraction. I cannot do better than refer to the Paper for the arguments upon which his conclusions are founded, which will be found published in the *Medical Times and Gazette*, June, 1855.

than with organic substances; the essential condition
in all these experiments being, that the current
should traverse the nerve, the *effect*—the muscular
contraction—remaining the same, whether the current had its origin in the animal body, or from
arrangements formed external to the body, and independent of it.   VOLTA's experiments appear to
have been adapted for the purpose of ascertaining
the effect of *current* electricity upon the animal body;
GALVANI's, for the purpose of ascertaining the *origin*
of the effects that were observed to occur in the
animal; and whatever views he might have entertained in regard to the nerve force being *identical*
with ordinary electricity, all subsequent experiments
only tend to shew the great difference that exists
between these two agents,—the action of current
electricity, and that of nerve force.   In saying this,
however, I am very far from denying that *nerve force*
is a *polar* force, and consequently it must bear some
relation or *connection* with the other *polar* forces;
nerve force being, as DR. TODD has pointed out, a
higher *form* of polar force, and perhaps the *highest
form* that we are acquainted with; and being such,
we may reasonably suppose that its chief and perhaps
peculiar *polar* characteristics can become manifested
in the animal body alone.   In another chapter this
question will again come under consideration.

The following conclusions in regard to the *origin*
of the so-called muscular and nerve currents may
be deduced from the foregoing experiments:

First, They may depend upon the changes which occur *during* nutrition.

Secondly, They may depend, also, upon the *heterogeneity* of the parts; (including under this term the action of the platinum electrodes, viz. the *catalytic* action or *combining* power of platinum upon the moist animal substances in contact with its surface ;) and also,

Thirdly, Upon a *polarized* condition of the nerve or muscular fibre.

Fourthly, That this *polarized* condition of the fibre is produced and maintained by *nutrition*. And,

Fifthly, That the circumstances under which the current may be produced may resemble, in one instance, those arising from the changes which occur in the decomposing cell of a voltaic circle; and in the other, those that arise from the action of a charged Leyden jar. In the latter case, the current being due to the *polarized* condition of the fibre; in the former, to the *changes* which occur *during* nutrition.

CHAP. IX.

ON THE POLARIZED CONDITION OF THE MUSCULAR AND
OF THE NERVOUS TISSUE IN THE LIVING OR RECENTLY
KILLED ANIMAL. MUSCULAR FORCE AND NERVE
FORCE, POLAR FORCES.

HAVING arrived at the conclusion, in the last
chapter, that the muscular and nervous tissues are,
during life, in a peculiar electrical state or condition,
and which has been termed *polarized*, the following
question naturally arises, Can this state, dependent
as it evidently is upon *nutrition*, be *increased* by any
artificial means? That it may be diminished or
easily destroyed is to be inferred from the fact, that
whatever interferes with the proper nutrition of a
muscle or nerve, or disorganizes their structure,
whether by mechanical or chemical agencies, destroys
also the conditions upon which the *existence* of the
muscular or nerve currents depends; and it is, it
may be observed, from the manifestation of these
currents that the existence of this *polarized* condition is inferred. It is reasonable, therefore, to
suppose, that it might be by the employment of the
electric force (or *current*) that we should perhaps
obtain some evidence to assist in solving this
problem.

In considering the question of the influence of
electricity upon the muscular and nervous tissues,

we are necessarily brought to the examination of the various experiments that have been undertaken from the period of GALVANI's celebrated discovery up to the present time, in which *muscular contraction* has been induced by means of the agency of electricity upon the nerves. On the present occasion, however, as it is by means of the *galvanometer* rather than by *muscular contraction* alone that the evidence proper for the solution of our question might be obtained, it will not be necessary to enter into a critical review of the various results made known to us by our predecessors; but I shall state those facts in the form of a few propositions which appear to have been well-established by the labours of VOLTA, MARIANINI, NOBILI, MATTEUCCI, MARSHALL HALL, DU BOIS REYMOND, and others[a], and to which I shall have occasion to allude in the course of the inquiry.

1*st*, When an electric current traverses a nerve, it is only at the *opening* and *closing* of the circuit that *muscular contraction* ensues.

2*dly*, No *muscular contraction* occurs *during* the passage of the current.

3*dly*, After the *inverse* current has passed for some time along a nerve, upon opening the circuit, *tetanic contractions* are produced; with the *direct* current *no tetanic contractions* take place.

[a] In BECQUEREL's Traité de l'Electricité, will be found an account of the views of the earlier inquirers, and also some very valuable observations of his own in regard to animal electricity. I may also refer to DE LA RIVE's Treatise on Electricity, translated by Walker.

The question may now arise, Can the *muscular contraction* of a limb be considered as evidence of an *increase of the polarized* state or condition of the nerve going to that limb? Previous to considering this question, let us endeavour to ascertain whether any increase occurs in the muscular or the nerve current under these circumstances.

MARIANINI[b], under the supposition that electricity, in these experiments, accumulated in the tissues, says, " En appliquant avec soin les fils du galvanomètre aux fibres palpitantes ou aux nerfs adhérens, on pourrait, peut-être, détourner en partie ces courans, et les faire passer par le galvanomètre; mais les expériences que j'ai faites jusqu'à présent sur ce point aussi délicat ne me permettent pas encore de rien affirmer avec assurance."

MATTEUCCI[c] adds, " How is this tetanic action produced? It is easy to convince one's self, if any doubt could be entertained upon the subject, that there is no electricity rendered latent either in the nerves or in the muscles by the passage of the inverse current. My endeavours to discover signs of any, by the aid of the condenser, have been entirely fruitless. Likewise there are no signs, on opening the circuit, of any electric current in circulation. I have made myself quite certain of this fact by means of the galvanometer, employing at the same time a pile of tetanized frogs."

[b] Annales de Chimie et de Physique, tom. lvi. p. 387, 1834.
[c] Phil. Trans. 1848, part ii. p. 236.

In the following experiments, the current either from 1, 3, or 6 of Grove's middling-size cells was passed through a detached muscle, or a portion of the sciatic nerve, the platinum electrodes being so arranged, that the *anode* or *platinum* extremity was in contact with the *base* or *transverse* section of the fibre, and the *cathode* or *zinc* extremity in contact with the *external* or *longitudinal* surface ; the *direction* of the current being in accordance with the normal direction of the muscular or nerve current, or that which is called the *inverse* current. The muscle or nerve, which was from rabbits, guinea-pigs, and frogs, was placed upon two pieces of glass, separated from each other, so that the current should traverse the substance of the tissue ; but how far it was conducted by the surface and not by the fibre alone, may be a question difficult to decide.

The normal effect of the muscular or nerve current was first ascertained and noted ; the former amounting to 4° or 5°, the latter to 2° or 3°, depending, however, upon the state of the nerve or muscle, and also of the animal. The current from the battery was allowed to pass along the fibre for different periods of time, when the effect upon the muscular or nerve current was then examined.

The effects, generally speaking, were as follow :

With the current from *one* cell, after it had passed five minutes, the amount both of the muscular and nerve current was slightly diminished ; after ten minutes, the nerve current was not obtained, the

muscular current still existed; fifteen minutes, no
nerve current, nerve dry, muscular current but slight;
twenty minutes, no effect; twenty-five minutes, no
effect.

With the current from *three* cells, after five
minutes, the nerve current very slight, the muscular
current somewhat diminished; after ten minutes, no
nerve current, nerve dry, muscular current slight;
fifteen minutes, no effect from nerve, muscular
current very slight; twenty minutes, no effect from
either; twenty-five minutes, no effect.

With the current from *six* cells, after five minutes,
no nerve current, nerve dry, muscular current very
slight; after ten minutes, indefinite indications of
the needle with the nerve, no muscular current;
fifteen minutes, no effect with nerve, with the
muscle the current occasionally indicated a reversed
direction; twenty minutes, no effect with the nerve,
indefinite indications of the needle with the muscle;
twenty-five minutes, effects similar to those last
observed.

As no distinct evidence of an *increase* either in the
muscular or nerve current was obtained, it has not
been thought necessary to particularize the results.
Great care was taken in all these experiments to
depolarize the platinum electrodes after the com-
pletion of each circuit, and to have them well
cleaned.

In the next series of experiments, the current was
passed *direct*, so as to traverse the fibre in the con-

trary direction to that of the muscular or nerve current. As the results did not indicate any *increase*, or even any decided *diminution* in the muscular or nerve current, and were in many respects similar to those I have already related with the *inverse* current, they need not be detailed.

Employing litmus paper to test the indications of the two surfaces of the fibre in contact with the electrodes, the effects were the same; there was no distinct acid or alkaline deposit on either of the two surfaces. The tissues soon became dry.

The results in all these experiments, after the current had passed for some time through the nerve or muscle, and especially when more than one cell was employed, evidently arose from the electro-chemical actions set up in the moist animal substances, by means of the electric current employed destroying the conditions necessary for the existence of the muscular or nerve current, and setting up new actions. It is extremely doubtful whether the current *traversed* the fibre[d]; it might have been conducted by the surface alone; and, in considering the action of the current upon the tissue, it is not difficult to perceive that its effect (its mode of action) would be not that of *increasing* the electrical state of the tissues, but more of a *disorganizing* action. What is evidently required is, to *induce* the *negative* electrical state of the fibre, which constitutes its *polarized*

---

[d] MATTEUCCI believes that the current is conducted by the liquid part of the nerve. Phil. Trans. 1850, p. i. p. 388.

condition; and it does not appear that, by merely passing a current through it, we should be able to produce the effect we wish.

Instead of a *constant*, an *intermitting* current, from an ordinary medical electro-magnetic machine, was employed, and made to traverse the muscular or nerve fibre as before. In these instances, there was no indication of an *increase* in either of the muscular or nerve currents.

The limbs of a galvanoscopic frog were now placed in separate glass vessels, employing the current from the battery of six cells, the current being *inverse* in one limb and *direct* in the other, as in MATTEUCCI'S experiment. When *tetanic contractions* were produced in the *inverse* limb, an attempt was made to ascertain whether any difference existed between the muscular and nerve currents of both limbs. No decided difference between the two limbs could be detected; differences were occasionally observed, but the nerve current in the *direct* limb was apparently as frequently *increased* as that of the *inverse* limb, but the muscular and nerve currents of both limbs, and in other parts of the same frog, generally indicated a greater amount of deflection previous to the passage of the current from the battery than afterwards.

It may be asked, Do not these *tetanic contractions* indicate an *increase* in the polarized state or condition of the nerve? Du Bois Reymond[e] states,

---

[e] On Animal Electricity. Edited by H. Bence Jones, M.D. p. 213.

that he has obtained indications of an *increase* in the nerve current, by passing an electric current along a portion of the same nerve.   He says, " If any part of a nerve is submitted to the action of a permanent current, the nerve, in its whole extent, suddenly undergoes a material change in its internal constitution, which disappears on breaking the circuit as suddenly as it came on.   This change, which is called the electro-tonic state, is evidenced by a new electro-motive power, which every point of the whole length of the nerve acquires during the passage of the current, so as to produce, in addition to the nerve current, a current in the direction of the extrinsic current.   As regards this new mode of action, the nerve may be compared to a voltaic pile, and the transverse section loses its essential import. Hence the electric effects of the nerve, when in the electro-tonic state, may also be observed in nerves without previously dividing them."

The experiment was repeated in the following manner :—Platinum wires, connected with the galvanometer, were placed in contact with the two portions of the nerve, (the *longitudinal* and the *transverse* sections,) so as to obtain the nerve current. The current from one, three, or six of Grove's middling-sized cells was then passed along another portion of the nerve, at different distances from that portion connected with the galvanometer, the electrodes of the battery being pointed.   When the current from the battery was confined to a small portion of the nerve, and at some distance (an inch or more) from the other portion, it very rarely happened that I could obtain any effect upon the nerve current. When the electrodes of the galvanometer comprised half-an-inch of the nerve, and the electrodes of the battery also the same extent, and were not far from those connected with the galvanometer, then an effect was frequently produced upon the nerve current; the effect, however, was, generally speaking, that of a *decrease* in the nerve current, and this took place, whatever might be the direction of the current from the battery, whether coinciding with, or in opposition to, the direction of the nerve current. It very seldom happened that an *increase* in the nerve current was obtained; and, as these results were chiefly observed to occur when more than one cell was employed, the effects—the apparent *increase*, and perhaps the *decrease*, of the nerve currrent— may be more correctly referred to some disturbance in the position of that portion of the nerve between the electrodes of the galvanometer, arising during the passage of the current in the other portion; an *increase* occurring when the nerve pressed against the electrodes, and a *decrease* when it separated from them. It may be just remarked, that the nerves soon became dry in these experiments.

The nerves were taken from the frog, guinea-pig, and rabbit.

Similar results were obtained when muscles

were employed in the same manner for the purpose of ascertaining the effect upon the muscular
current.

Although the effects observed may partly arise
from an alteration in the contacts of the two surfaces
of the nerve between the electrodes of the galvanometer, I nevertheless believe, with Du Bois
Reymond, that the passage of an electric current in
another portion of the same nerve is capable of
affecting the conditions upon which the nerve
current depends; but I have not been so fortunate
as Du Bois Reymond in obtaining such decided and
definite evidence as could have been wished in regard
to the *increase* of the nerve current.. In all these
experiments, the *distance* of the electrodes, both of
the galvanometer and battery, from each other, and
the *extent* of the nerve between each of the electrodes, and also the state of the nerve in regard to
the dryness of its surface, are points of the utmost
importance to be considered and attended to in
judging of the final result.

' The only conclusions that can be deduced from
the foregoing investigations are the following:

' 1st. That we have no evidence of being able to
increase the *polarized* condition of the nervous and of
the muscular tissue by artificial means, such as the
electric current; but it is highly probable,

2nd, That an *increase* of this *polarized* condition
may arise from an increased action of those changes
which take place in the living animal, such as

nutrition, being the same means by which it is produced and maintained.

Before acceding to these conclusions, it may be reasonably asked, Have we not other evidence, besides that afforded by means of the galvanometer, to indicate an *increase* in the polarized condition of the nerve? Do not the *tetanic contractions*, which are observed in a limb whose nerve has been subjected to the action of an electric current (*inverse*), indicate an *increased* action of the nerve?

A current from six of Grove's cells was passed through the limb of a galvanoscopic frog in the *inverse* direction, and as soon as *tetanic contractions* were produced, the nerve was divided at the junction of the nerve with the muscles of the limb; the tetanic contractions ceased. The two ends of the divided nerve were now placed in apposition, but no tetanic contractions ensued. This *inverse* current was again allowed to pass for some time through the nerve thus united, but no tetanic contractions occurred upon the breaking of the circuit. Great care, however, is required in this experiment to divide the nerve at the exact point where it emerges from the muscles, as pointed out by MATTEUCCI, otherwise the tetanic contractions take place.

The results of this experiment only tend to confirm what has been already satisfactorily proved by others, that the *continuity* of the nerve fibre in the nerve leading to the muscle is necessary for the conduction of the impression excited at the distal end of the nerve, in order to arouse muscular contraction. It need scarcely be added, that the muscular and nerve currents may, however, be obtained under these circumstances between the separated portions.

The results of these investigations lead to the conclusion, that the muscular and the nervous tissues are both, during life, in a *polarized* state or condition, and from our inability to *increase* this state by any artificial means, it being produced and maintained by *nutrition*, would almost stamp it as being peculiar to the organic kingdom. All the experiments tend, moreover, to shew, that it must be by acting upon and employing the means by which it is produced and maintained, viz. through the act of nutrition, that we can hope to succeed; and it is reasonable to suppose, that whatever would increase this act would also increase this condition, as shewn by an increase in the muscular and nerve currents under these circumstances. That *tetanic contractions* may be produced by means of the electric current upon the nerve, might perhaps be adduced as an argument in favour of the supposition that an increased action of the nerve is produced, and, consequently, an *increase* in the polarized condition; but how far the peculiar state of the nerve which produces *tetanic constructions* under these circumstances is due to an *increase* of the normal *polarized* condition is questionable, as I shall endeavour to point out.

When an electric current from a voltaic circle or

any other source traverses a nerve or muscle, the muscle or nerve forms a part of the conductor, as any other moist substance or fluid (electrolyte) might do, and becomes polarized; the polarized condition of the muscle or nerve under these circumstances being manifested, as the other points of the circuit, in the *longitudinal* direction, one extremity being positive or negative to the other; and when the electrodes of a galvanometer are applied, one to each end, an electric current is manifested. Nothing of this sort, however, is manifested in the normal polarized condition of the muscular or nerve fibre as it exists in the animal body; the polar action, as then manifested by the galvanometer, is in the *transverse* direction, or, more correctly speaking, the tissue is in one electric state *negative*, and the *positive* state is external to it either in the sarcolemma or neurilemma, or in the blood. Hence we see the important difference between these two modes or conditions or states. To suppose, therefore, that an electric current, in traversing a muscle or nerve, increases its normal polarized condition, would only lead to erroneous ideas of the subject; it may, however, induce a change, under certain circumstances, in the condition of the nerve, tantamount to an increased action of the nerve or nerve force; but it may be very much doubted, without knowing what this change is, whether we should be justified in calling it an *increase* of its natural *polarized* condition.

Since, then, the nerve fibre and the muscular

fibre present this peculiar polarized condition, we may, without any hesitation, infer, that the two forces, the muscular and nerve force, are both POLAR, and, consequently, that muscular action and nerve action are *polar* actions. With regard, however, to the nerve, the mere circumstance of its electrical state will not alone prove that nerve force is polar, as we shall see in a subsequent chapter.

During life, the muscular and nerve fibre may be considered as existing in a state of *tension*[f], a *forced* state, and muscular relaxation may coincide, and would be synonymous, with this polarized condition or state of *tension,* and *contraction,* the result of a *depolarization* of the muscular fibre. In the normal state, this *depolarization* is induced by means of nervous agency; but it may also occur from other means, such as chemical or mechanical agencies, or whatever is likely to produce disorganization. *Depolarization* having occurred, the polarized state is easily and readily restored by the act of *nutrition;* hence *contraction* may be partly considered as the result of *molecular attraction*[g]. The particles con-

---

[f] It would be of some importance to ascertain the state or condition of the prisms in the electric organ of the fish, prior to their discharge of electricity. Theoretically, all the tissues in the animal body may be considered as being in an electric or polarized state, presenting, however, great differences in regard to each other.

[g] I have already alluded to the opinion of Dr. Radcliffe, who considers that muscular contraction is the result of molecular attraction.

stituting the muscular fibre being in a state of self-repulsion whilst in this polarized condition, would resist the force of molecular attraction unless depolarized; and, according to this view, the results of muscular contraction would necessarily follow in conformity to the law as laid down by SCHWANN.

The nerve fibre existing also in a state of *tension*, whatever influences this state would also produce a *depolarization* of its fibre, the result being manifested by muscular contraction, or pain. As it is only during the period of this *depolarization*, when the *tension* is altered, that contraction occurs, we have some clue, as pointed out by DR. TODD[h], to the reason why contraction only ensues at the *opening* or *closing* of the circuit, and not *during* the continuance of the passage of the current along the nerve. The constant current does not induce those *momentary* changes necessary for the production of muscular contraction; it is only at the *opening* and *closing* of the circuit that the *tension* of the fibre is affected. The question, however, may be asked, How de we account for the fact, that an *inverse* current through a nerve will cause *tetanic contractions*, whereas a *direct* current has no such effect? In this instance, the *inverse* current may produce an *altered* condition of the *polarized* state; but I do not believe, as has been already stated, that it *increases* the normal polarized condition; and if contraction be due to momentary

[h] Cyclopædia of Anatomy and Physiology; *Art.* " Physiology of the Nervous System."

changes in the condition of the nerve, the tetanized state of the muscle might be a necessary consequence until the nerve had acquired its normal and permanent condition. But the same remarks might be made of the nerve in which the *direct* current has passed, and yet no contraction occurs. This is a difficulty which at present we cannot satisfactorily overcome; but the fact is of some importance, as indicating a dependence upon the *direction* of the electric current in its passage along the nerve, and shewing that a peculiar state is induced in one case and not in the other[i]. My endeavours to ascertain what this state may be by means of the galvanometer have hitherto failed.

In all these discussions, the importance of considering the conditions both of the nerve tissue and muscular tissue, in regard to their *nutrition*[k], cannot

[i] MATTEUCCI believes that the *excitability* of the nerve is increased by the *inverse* current, and diminished by the *direct* current.   Phil. Trans. 1846, part iv. p. 483. It is possible that the *inverse* current may keep up the normal electric condition of the nerve, inasmuch as the current is similar in direction to the natural current, and the *direct* current may destroy this normal state, being in opposition to it.   Under these circumstances, the *nerve current*, if not *increased* by the *inverse* current, should indicate some evidence of its *continuing* longer, and the opposite effect should be produced by the *direct* current.   At present my experiments have failed to shew this.

[k] I cannot do better than refer to the article on the " Physiology of the Nervous System" in the Cyclopædia of Anatomy and Physiology, by Dr. Todd, in which this fact is clearly pointed out.

be too strongly insisted upon. It is by nutrition that the *force* is maintained, and it is by this act that it is renewed. If the changes in each tissue are not properly carried on, or should fail in one or both of them, spasms or convulsions may be induced at one time, or the contrary effect, such as paralysis, might occur at another.

It may be urged as an argument against the views now taken, of *muscular* force and *nerve* force being *polar*, that they are considered as identical. So far as they are POLAR they are identical; but the conditions under which the force exists in the two tissues differ. In the muscle it is limited to, and associated with, that tissue, and may be considered as existing in a *static* form. In the nerve it may also exist in this form, the *static*, but it is capable also of being *transmitted* from one point to the other, in consequence of the peculiar organization of the nervous tissue. The identity between the two forces removes those difficulties that might otherwise arise in considering their reaction one upon the other. *Muscular force* and *nerve force* must not be supposed to be *produced* by their respective tissues, but *associated* with them; the *modes* under which they are manifested differ, and would indicate that the *form* under which *nerve force* exists would prove it to be of a higher character than that of *muscular force* [1].

---

[1] *Vide* FARADAY's Paper on the Gymnotus, (Experimental Researches, vol. ii. p. 16.) on the " Relation between Nervous Power and Electricity."

It may be urged, if *nerve force* be a *polar force*, how is it possible to conceive that mental acts, that phenomena connected with the *mind*, should be the result or the manifestation of *polar action?* Without entering into the discussion as to how far mental acts are the *result* of an action of the brain, or whether it be through the *medium* of the brain that the mind acts upon the body, or whether mental phenomena are *polar* phenomena, I nevertheless believe, that the phenomena associated and connected with nervous action are confined to, and can only become manifested in, the animal kingdom; and although they may bear a close relation or *connection* with the action of other *polar forces*, they will nevertheless form a distinct class, a class *sui generis*, and perhaps the highest *form* by which *polar* phenomena can become manifested.

ON THE MANIFESTATION OF ELECTRIC FORCE, OF
CHEMICAL FORCE, AND OF HEAT, DURING MUSCULAR
CONTRACTION.

SECT. I. *On the Manifestation of Electric Force during
Muscular Contraction.*

I shall purposely abstain from entering upon the
personal disputes that have arisen during the discussion of some of the questions that will now
occupy our attention, and only refer in a general
note to the principal sources from whence I have
been enabled to obtain such views as are confirmed
by experimental evidence, and which are now
entertained upon the subject of our present inquiry[a].

[a] Annales de Chimie et de Physique, 3e série, vol. xv. p. 64;
vol. xxx. pp. 119, 179; vol. xxxix. p. 114. Comptes Rendus,
vol. xxviii. pp. 570, 641, 653, 663, 782; vol. xxx. pp. 349, 406,
479, 512, 563, 699; vol. xxxi. pp. 28, 91, 318; vol. xxxii. p. 131.
Bibliothèque Universelle de Genève, Fev. 1850, Juin 1853.
On Animal Electricity; being an Abstract of the Discoveries of
Emil du Bois-Reymond, &c. edited by H. Bence Jones, M.D.
F.R.S. London: Churchill, 1852, 8vo. Lettre de Charles Mat-
teucci à M. H. Bence Jones, F.R.S. Florence: Imprimerie Le
Monnier, 1853, 8vo. On Signor Carlo Matteucci's Letter to
H. Bence Jones, M.D., F.R.S. &c. By Emil du Bois-Reymond.
London: Churchill, 1853, 8vo. As the original Paper was read
before the Royal Society in 1855, my readers will find some
further observations of MATTEUCCI in regard to the subject in

The fact long since obtained by MATTEUCCI, viz. that of causing the muscles of a galvanoscopic frog to contract by placing its nerve upon the muscles of another animal during their contraction, is well known; and in a communication[b] to the Royal Society, MATTEUCCI has adduced strong reasons for believing that an electric *disequilibrium* is produced during muscular contraction.

DU BOIS REYMOND[c] has also satisfactorily shewn, that when the hands are placed in two separate vessels containing water, and connected with a galvanometer, that then, upon making the muscles of one arm contract, an effect is produced upon the needle. These results have been confirmed by ZANTEDESCHI[d]. BUFF[e], and TYNDALL[f].

The results obtained by DESPRETZ[g], BECQUEREL[h], and MATTEUCCI[i], have thrown strong doubts upon the conclusions deduced by DU BOIS REYMOND. Several experiments undertaken by myself, both prior and subsequent to the knowledge of DU BOIS REYMOND's researches, tended to confirm in my

the Bibliothèque Universelle de Genève, in the Annales de Chimie et de Physique, and in the Comptes Rendus for the year 1856.

[b] Philosophical Transactions, 1850. Ninth Series.

[c] Comptes Rendus, vol. xxviii. p. 641.

[d] Bibliothèque Universelle de Genève, Fev. 1850.

[e] Ibid.

[f] British and Foreign Medical Review for Jan. 1854, p. 141.

[g] Comptes Rendus, vol. xxviii. p. 653.

[h] Ibid. p. 668.

[i] Ibid. p. 782.

opinion the doubts expressed by these inquirers. But the positive evidence adduced by Du Bois Reymond by means of his delicate galvanometer, combined with that of Matteucci obtained by means of the frog, only made it incumbent upon the opponents of Du Bois Reymond to disprove his conclusions, if capable of disproof, more by stronger experimental evidence than by theoretical arguments; hence the renewal of the present inquiry.

In relating the experiments, to avoid unnecessary prolixity, I shall endeavour to be as concise as possible, although it may be considered that the various sources of error have not been sufficiently eliminated. My great point, however, will be to establish facts; and whatever observations it may be necessary to make in reference to the *origin* of the effect produced, these will form the subject of the concluding remarks, and I shall just relate in a general manner some results that were first obtained by means of a galvanometer consisting of but few coils. To two brass handles of an ordinary medical electro-magnetic machine, were attached thick copper wires, each about 8 inches in length, and bent. Similar copper wires were connected with the galvanometer, their free ends communicating with mercurial cups, these cups forming the means of connexion with the brass handles. Holding the metallic handles firmly, grasping one of them and contracting the muscles of the arm powerfully, and doing this alternately with each arm, I at first failed to obtain any, or if any, but indecisive results. Upon repeating the experiments on another occasion, some definite result was ultimately obtained, and also a clue to my former failures. I found that when the muscles of the arms were contracted alternately at definite periods, and continuing this act for some time, that *as the hands became moist*, a decided effect upon the needle occurred, indicating the contracting arm to be *positive*[k] to the other. These results may be considered as due partly to the reaction of the acid secretions of the hand upon the metallic electrodes, partly to the skin becoming a better conductor than the dry cuticle, and partly perhaps to thermo-electric action.

The question now arose, could not the effect which might be due to muscular contraction coincide with that which arose from the action of the cutaneous secretions? From the results obtained by Du Bois

[k] Some difficulty is frequently experienced in comprehending the results obtained by different inquirers in reference to the *direction* of the current; it is uncertain to which of the circles of the old construction, the simple or compound, reference is made for illustration, or whether it be the *positive* metal or the *positive* electrode that is referred to; hence arises an apparent contradiction. The *direction* of the current is of the utmost importance to be attended to in accounting for the effects produced. I always allude to the simple elementary circle for illustration; if zinc, platinum, and dilute acid, be formed into a circuit, the current goes, according to the usual mode of expression, from the zinc *to* the platinum; the platinum being the *positive* electrode. In the *combination* of an acid with an alkali, the electrode in contact with the *acid* is the *positive* electrode.

REYMOND[1], it would appear, that the *direction* of the current due to muscular contraction is *inverse*, i. e. *from* the hand *to* the shoulder.

Several other experiments were undertaken, in which different solutions were employed to moisten the hands; the effect upon the needle was now much greater. When one hand was moistened with a weak alkaline solution and the other with water, upon contracting the muscles of the arm holding the alkaline electrode, this became *negative* to the other; but by proper management, by not having the solution too strong, this arm could be made to become *positive*, the current occurring during muscular contraction overcoming the influence of the alkaline solution. Feeling the force of the objections that might be raised in reference to the use of the metallic electrodes, I do not think it necessary to relate any other experiments in which these instruments were used.

A galvanometer[m] of the following construction was now employed. It consisted of two coils, placed one above the other so as to produce the full effect upon both needles, with an index to indicate the amount of deflection. Each coil was 2 inches in length, and the same in breadth, consisting of 1500

---

[1] Rapport sur les Mémoires rélatifs aux phénomènes électrophysiologiques présentés à l'Académie par M. E. du Bois-Reymond, Comptes Rendus, vol. xxxi. p. 28.

[m] Vide FARADAY's mode of employing a galvanometer. Phil. Trans. Series xxviii. 1852.

turns; the thickness of the wire 37 gauge. The needles were rather less than 2 inches in length; the index, of ivory, 3 inches in length. As my object was to ascertain, if possible, the *existence* of the force rather than the *amount*, a further description of its delicacy will not be necessary.

Two copper clamps were made, each 4 inches in length, tapering at one extremity so as to be connected with the binding-screws of the galvanometer, the other extremity being bent down at right angles to the extent of half an inch, presenting a surface 1¼ inch in breadth, and to which was also attached a piece of copper by means of two screws; by these the platinum electrodes were securely fastened, and those that were generally used consisted of platinum foil, each 2¼ inches in length and 1 inch in breadth.

The vessels usually employed to hold the solutions were two cupping-glasses, each 2¼ inches in diameter, 2¾ inches in depth, and contained rather more than four ounces of fluid. The glasses being half filled with a weak solution of common salt, covering the electrodes to the extent of an inch, two and sometimes three fingers of each hand were held perpendicularly in the vessels, the middle finger resting upon the bottom of the glass vessel. When first introduced, a slight tremulous motion of the needle was observed; upon taking the fingers out of one vessel and reintroducing them, and repeating this act, a slight effect appeared, at first definite, but this soon subsided. Similar effects were observed when the fingers of the other hand were used in a similar manner. The fingers last introduced were not always *positive* or *negative* to the others. Keeping the fingers in, then moving one of the fingers so as to make the fluid rise and fall upon the surface of the electrode in the vessel, produced but little motion of the needle. After the fingers had been in for some time, and the needle had become quite stationary, upon contracting the muscles of one arm an effect occurred upon the needle indicating the contracted arm to be *positive* to the other $3°$; by contracting the muscles of each arm alternately, the effect amounted to $5°$. The result was definite, and the motion of the needle steady.

Du Bois Reymond[n] considers that there are five circumstances as influential in producing a current: viz. 1*st*, température inégale; 2*nd*, durée inégale de l'humectation avec le liquide conducteur; 3*rd*, tension inégale de la peau; 4*th*, lésion de l'une des places de la peau; 5*th*, transpiration inégale. I could not refer the effects that were obtained to either of these circumstances; to the tension of the skin, for instance.

Three solutions were now prepared; the first consisting of a concentrated solution of common salt; the second of sulphuric acid, one part of strong acid to six parts of water; and the third of one part of liq. potassæ (Pharm. Lond.) to four parts of water.

[n] Bibliothèque Universelle de Genève, Juin 1863.

It was found necessary that the following pre-cautions should be strictly attended to. The solutions should be prepared and well mixed previous to their use, and the electrodes covered to the same extent. The hands and fingers should be thoroughly *clean*; to attain this object, they were first washed with the ordinary curd or Windsor soap, and afterwards well rinsed in common water, and wiped comparatively dry with a clean towel. The same towels should not be employed when using the different solutions, and it was found better to work on different days with each solution; for the liquids soaking in between the nails and the fingers became a source of some difficulty to remove, and a cause of some embarrassment in judging of the final result.

It need hardly be stated, that the surfaces of the electrodes should be clean and the contacts perfect.

With the solution of common salt. Upon the first introduction of the fingers, a slight tremulous motion of the needle occurred. Withdrawing the right-hand fingers and reintroducing them immediately after-wards, and repeating this act a few times, a slight *positive* effect was produced upon the introduction of the fingers. Withdrawing the left-hand fingers in the same manner, a slight *negative* effect was at first obtained, but after a short time the effect became indefinite, and ultimately slightly *positive*. Keeping the fingers in and contracting the muscles of the arm, the fingers of the contracted arm were *positive* 4° to 5°, and made to increase.

The solution in one vessel was now diluted to one-half. Upon the first introduction of the fingers, the strong solution was *positive* 5°. Keeping the fingers in and waiting until the needle had become motionless, the contracted arm was *positive* to the other; if the arm connected with the strong solution was contracted, the effect was greater than when the other was contracted, the current in the former case rising to 5°, in the latter to 2° or 3°.

With the acid solution. Upon the first introduction of the fingers, the needle would sometimes go 30° or 40°, but generally speaking it would oscillate first to one side and then to the other. Withdrawing the fingers of one hand and then reintroducing them, this hand was *positive* to the other. Keeping the fingers quiescent, and contracting the muscles of the arm, the corresponding fingers were *positive* 4°.

The solution in one vessel was diluted to half the strength of the former; the strong solution was *positive* to the other upon the introduction of the fingers. If the difference between the two solutions was not too great, (the exact strength being difficult to state,) I could determine a slight current to pass in a constant direction, and then make the current arising during muscular contraction overcome this constant current. Strong solutions were found best for this purpose, from their forming, perhaps, a better conducting liquid than the weaker solutions.

With the alkaline solution. Upon the first introduction of the fingers no definite result, the needle tremulous. Withdrawing the fingers and reintroducing them, the effects varied, but, generally speaking, the reintroduction of the fingers produced a *neyative* result. The effect due from muscular contraction was definite from 2° to 4°, the fingers of the contracted arm being *positive*.

The solution in one vessel was diluted to half the strength of the other. Upon the introduction of the fingers, the diluted solution was *positive* to the stronger. I was enabled in this instance, as with the acid solution, to obtain a solution of such strength as to give a constant current in a certain direction, and then make the current due from muscular contraction overpower this current. As this fact is one of some importance, it will be necessary to state the mode of ascertaining this constant current. If we find, either upon the introduction of the fingers of both hands at the same time, or upon the introduction of the fingers of one hand, that a current existed in one direction to the amount of about 2°, the current occurring during muscular contraction, if it coincided with this constant current, would cause the needle to advance to 4° or 5° or more; but if it has to overcome this constant current, the needle would only move 2° or 3°. I need scarcely add, that the needle of the galvanometer should continue to indicate this current whilst the fingers remain in the solution, which may be indicated by the needle receding to its normal position upon the withdrawal of the fingers. A great difficulty, however, is frequently experienced in obtaining this constant current.

Several other experiments might be related in which the electrodes were made to differ in size, one being narrow, the other broad; or where one was made to dip deeper into the solution than the other. The general results indicated that a difference in extent between the surfaces of the platinum electrodes and the solutions, occasioned an effect upon the needle upon the first introduction of the fingers. In other experiments larger vessels were employed, so as to insert the whole of the hand and wrist; the results coincided with those that have been related, the effect however was greater.

The solutions were made to vary as to temperature by placing the glass in a vessel containing hot water, and then carefully stirring the solution so as to make it completely uniform. One vessel was at the temperature 65° F. and the other 115°. The results upon the first introduction of the fingers were indecisive. The hot solution was not always positive to the cold solution. The effect during muscular contraction could always be obtained.

I agree with Du Bois Reymond as to the importance of attending to the five circumstances to which he refers, and which have been already alluded to. The circumstances that appear to me to be the most influential in masking the ultimate result are, 1st, the action of the fluid upon the surface of the electrodes; and, 2ndly, the reaction of the cutaneous secretions upon the fluids. Although we cannot remove these sources of error, we have it nevertheless in our power to counteract their effects, by shewing that the current consequent upon muscular con- traction can be made to overcome that arising from either of the two former actions.

I have not yet alluded to the difference which appears to exist between the results obtained by DU BOIS REYMOND and myself, viz. the *direction* of the current. DU BOIS REYMOND's experiments indi- cate that the current is *inverse*, i. e. *from* the hand *to* the shoulder during muscular contraction; my re- sults, on the other hand, point out that the current is *direct*, i. e. *from* the shoulder *to* the hand. I do not think that my results differ much from those of DU BOIS REYMOND; but as the discussion of this question will involve certain theoretical views, what- ever observations I may have to make will be deferred to the concluding remarks °.

° My original Paper was read before the Royal Society, since then I have had the opportunity of attending the interesting lectures of DU BOIS REYMOND at the Royal Institution. DU BOIS REYMOND considers, as far as I could understand, that when the whole of the hands are immersed in the solution, and the muscles of the arm are made to contract, then the current is due to the swollen state of the skin, and perhaps to other cir- cumstances; but should we be justified, I would ask, in con- cluding, that when the *fingers alone* are in the vessel, that then the effect is due to their swollen state? If DU BOIS REYMOND maintains that the current due to muscular contraction is *inverse*, i. e. *from* the hand *to* the shoulder, I can only add that I have never been able to obtain that constant effect. It cannot

With the assistance of three friends, an attempt was made to increase the effect upon the needle by forming a pile, as it were; we could obtain the effect, but there was no decided increase.

Several experiments were undertaken with the *rheoscopic frog*[p] in lieu of the galvanometer; the results were not so satisfactory as could be desired, to justify their being recorded.

———

The posterior limbs of a frog, separated at the pelvis, but connected by means of the lumbar nerves and a portion of the vertebral column, were each placed in separate vessels containing the solution of common salt; the muscles of one limb were then made to contract, and an immediate effect upon the

be denied that it frequently happens, upon the first repetition of the experiment, that the current may appear in favour of this opinion. DESPRETZ has remarked, that the current may appear first in one direction and then in another. I cannot insist too strongly upon the necessity of having the hands perfectly clean, and I am convinced that the failures and contradictory results which arise are due more to the want of attention on this point than to any thing else. The *direction* of the current is a fact of the utmost importance, as we shall hereafter see.

[p] The term *rheoscopic* has been recommended in the Report of the Committee of the Academy of Paris, in preference to that of *galvanoscopic*. I have employed both terms. When it is used for the detection of the *current*, the former term is most applicable; but the frog *may* be a test of a force in which the *current* force, in accordance with our present notions of force, cannot be shewn to exist. If the *dynamic* condition and the *current* condition of force be considered as equivalent terms, then *rheoscopic* would be unobjectionable.

needle was observed, the contracted limb being *positive* to the other from 3° to 4°. To prevent the fluid from being scattered upon the electrode, a piece of glass was attached to the part as a weight to avoid too great motion of the limb. In some instances both feet were removed : the effect upon the needle was still the same.

Similar results were obtained with the *acid* solution, but with the *alkaline* solution the effects varied, especially so if the solution were strong. The alkali acted apparently upon the mucous secretion of the skin.

Great care was required to have the cutaneous surfaces clean previous to the introduction of the limbs into the solutions, so that no current could arise therefrom.

Is it necessary to place the nerve of a galvanoscopic frog in contact with the longitudinal and transverse section of a muscle in order to obtain the necessary contractions? Repeating the experiment of MAT-TEUCCI[q], it was found that the contraction might be obtained without placing the nerve in contact with the muscle. At the same time I must add, that the effect upon the galvanoscopic frog appeared greater when its nerve was placed transverse to the muscular fibres than in any other position.

———

[q] Phil. Trans. 1850. Ninth Series. Annales de Chimie et de Physique, Juin, 1856.

The results of these experiments tend to establish the following conclusions, viz.

First. That during muscular contraction in man and in frogs, an effect upon the galvanometer may be obtained indicating the manifestation of an electric current.

Secondly. That this manifestation of an electric current is due, in a great measure, to secondary reactions, viz. between the animal secretions and the solutions on the one hand, and between the solutions and the platinum electrodes on the other; but that there nevertheless · remains a *residual effect* which cannot be referred to either of these actions, or to those pointed out by Du Bois Reymond.

Concluding Remarks.

The numerous instances in which an electric current may be shewn to exist, naturally renders any conclusion deduced from results obtained by means of the galvanometer extremely doubtful. If these remarks are applicable to physical researches, where we have such control over our experiments as to afford us facilities favourable for eliminating sources of error, how easily may we conceive that they would apply with far greater force to physiological researches devoid of such facilities. The strong prejudices which exist in reference to electro-physiological inquiries cannot therefore be a matter of much surprise, but must create a great difficulty to any individual who shall attempt the elucidation

of any electro-physiological problem experimentally;
for not only has he to satisfy, from the mixed character of the inquiry, the extreme views of the
physicist on the one hand, and those of the physiologist on the other, but they also afford the
indolent inquirer a ready means of apparent refu-
tation, and a powerful weapon to the controversialist.
That these opinions are not ill-founded might be
easily shewn[r]. Feeling the importance of basing
the conclusions in these inquiries upon strictly
experimental evidence, in discussing the different
views that may be entertained in regard to the
evidence upon which the experiments justify us in
concluding that some force *is* evolved during muscular contraction, I shall confine my observations to
the experimental results obtained by MATTEUCCI and
DU BOIS REYMOND, and endeavour to avoid every
thing of a purely controversial character.

MATTEUCCI[s] has proved, and shewn by means of
the galvanoscopic frog, that during muscular contraction the muscles of the galvanoscopic frog may
be excited to contract, and that for this purpose it is

---

[r] Strange to say, it has been urged, that the *time* has not yet
arrived for the prosecution of these experiments. This is the
only tangible objection that the author has yet had an opportunity of refuting. That objections may be started he is perfectly convinced; but as the inquiry professes to be *experimental*,
the objections must be supported by experimental evidence
before he can notice them.

[s] Phil. Trans. 1850. Ninth Series. Annales de Chimie et de
Physique, Juin, 1856.

not necessary that the nerve should be brought into contact with *different* portions of the muscular fibre. In repeating these experiments, I have been able to confirm the results obtained by MATTEUCCI.

MATTEUCCI[t], in his endeavours to ascertain whether the muscular current was increased during muscular contraction by forming piles with muscular elements, failed to obtain any decisive result. DU BOIS REYMOND, by means of his galvanometer, has satisfactorily shewn, that when the electrodes are so arranged with a muscular element that the current (muscular) passes through the instrument, that then, upon the contraction of the muscle, the needle of the instrument recedes, and this he calls " the negative variation" of the muscular current. This fact, established by DU BOIS REYMOND, and also the results obtained by MATTEUCCI, have been confirmed by some of my own experiments. But I cannot help thinking with MATTEUCCI[a], that the falling or the receding of the needle in DU BOIS REYMOND's experiment may be attributed in a great measure to the variation in the contacts between the surfaces of the muscle and the electrodes during the act of contraction. I am not aware that the existence of the muscular current is disputed by any one; and with respect to its *origin*, there can be no doubt that it arises from nutrition, and may be manifested under two points of view, as I have endeavoured to shew in chapters vi. and viii.

[t] Phil. Trans. 1845.
[a] Phil. Trans. 1850. Ninth Series.

We have then now two distinct questions for our consideration : 1st, Is there, or is there not, any force evolved during muscular contraction? and, 2ndly, Is the muscular current affected during muscular contraction? And this brings me to speak of the results we have been considering in the present chapter.

Du Bois Reymond's results lead him to consider, that in the frog the current is direct, and in man inverse, as shewn by the galvanometer. My experiments lead me to believe, that the current is direct in both instances. Upon carefully looking over a Paper by Du Bois Reymond* upon this question, it appears to me that he frequently obtains the direct current; but laying so much stress, as he does, upon the necessity of a current (muscular) circulating through the instrument prior to contraction before he can conclude that the resulting action upon the needle is what he calls "the negative oscillation" of the muscular current, that his object was to ascertain whether the muscular current in the human subject was not affected during muscular contraction, as in the frog, when arranged with a muscular element in the manner I have stated in the preceding paragraph. But then the question will arise, how can we account for the effects, when the limbs of a frog are placed in separate vessels, as is done with the hands? I may be mistaken in my opinion. As I have not been able to obtain this inverse current, and if Du Bois Reymond has been endeavouring to obtain this inverse

* Bibliothèque Universelle de Genève, Juin 1853.

current to prove " the negative variation" of the muscular current in the human subject, then the object that each of us has had in view differs: mine has been to ascertain whether any force *is* evolved during muscular contraction; DU BOIS REYMOND's, whether the muscular current is affected *during* muscular contraction. The importance of DU BOIS REYMOND's researches cannot be denied; but it appears to be, and is, of extreme importance that the two questions should be kept perfectly distinct, although perhaps intimately connected with each other.

The results of my inquiries lead me to believe, that during muscular contraction a force *is* evolved, as in the fish, but that it is only during extraordinary muscular exertion that it can become manifest to the galvanometer. I am perfectly aware of the objections that may be urged in reference to the fish being provided with a special apparatus; and in my first endeavours to obtain some result with the galvanometer, in which I failed to obtain any evidence of the existence of a force being evolved, I was then led to the conclusion as to the improbability of any force becoming free, as it were, during muscular contraction; that whatever force might be evolved would be expended or converted during that act; but I could never get over the difficulty which the results of MATTEUCCI, obtained by means of the frog, presented for consideration, and which is doubly increased by the results that have now been related in the present chapter. In reference to the fish, also, it must be remembered, that the force evolved bears some relation, according to MATTEUCCI[y], " to the activity of the functions of circulation and of respiration, and of every act of nutrition." The apparatus may be a *means* for the evolution of the force, but not a *producer* of the force; and there are some reasons for believing that the electric condition of the blood in the living animal must not be overlooked.

The question in what way the nerves influence muscular contraction, and have in the development of the force during muscular contraction, or in the evolution of the force in the fish, I am not now discussing. It is of primary importance to ascertain first of all *what* does take place. The influence that the nervous system has over these results is a subject for future consideration.

SECT. II. *On the Manifestation of Chemical Action during Muscular Contraction.*

That chemical changes take place in muscles *during* muscular action, has been inferred from the changes produced in the muscular tissue after violent exercise. MATTEUCCI[z], however, has shewn, that during muscular contraction oxygen is absorbed, and that carbonic acid and nitrogen are evolved. The mode of conducting the experiment is as follows:

[y] Phil. Trans. 1847, p. 241.
[z] Philosophical Magazine, June, 1856.

" Take," he says, " two wide-mouthed glass phials of
equal size, 100 or 120 cub. centims.; pour 10 cub.
centims. of lime-water (eau de chaux) into each of
these phials. Prepare ten frogs in the manner of
GALVANI; that is, reducing them to a piece of spinal
marrow, thighs and legs without the paws, which
are cut in order to avoid contact with the liquid in
the phials. The cork of one of these phials is
provided with five hooks, either of copper or iron,
on which five of the prepared frogs are fixed.
Through the cork of the other phial are passed two
iron wires, bent horizontally in the interior of the
phial; the other five frogs are fixed by the spinal
marrow to these wires. This preparation must be
accomplished as rapidly as possible, and both the
phials be ready at the same instant, and great care
taken to avoid the contact of the frogs with the sides
of the phials or the liquid. When all is in readiness,
with a pile of two or three elements of Grove, and
with an electro-magnetic machine such as is employed for medical purposes, the five frogs suspended
on the two iron wires are made to contract. After
the lapse of five or six minutes, during which time
the passage of the current has been interrupted at
intervals in order to keep up the force of the contractions, agitate gently the liquid, withdraw the
frogs, close rapidly the phials, and agitate the liquid
again. You will then see that the lime-water contained in the phial in which the frogs were contracted
is much whiter and more turbid than the same liquid

contained in the other phial in which the frogs were left in repose. It is almost superfluous to add, that I made the complete analysis of the air in contact with the frogs according to the methods generally employed." He calls the phenomenon, '*muscular respiration*,' and states, that "a muscle which contracts, absorbs, while in contraction, a much greater quantity of oxygen, and exhales a much greater quantity of carbonic acid and azote, than does the same muscle in a state of repose. A part of the carbonic acid exhales in the air, the muscle imbibes the other part, which puts a stop to successive respiration, and produces *asphyxy of the muscle*. Thus a muscle even ceases to contract under the influence of an electro-magnetic machine enclosed in a small space of air: this cessation takes place after a longer interval of time if the muscle is in the open air, and much more slowly still if there be a solution of potash at the bottom of the recipient in which the muscle is suspended. Muscles which have been kept long in vacuum or in hydrogen are nevertheless capable, though in a less degree, of exhaling carbonic acid while in contraction; this proves clearly, that the oxygen which furnishes the carbonic acid exists in the muscle in a state of combination."

That different gases present great differences in their influence upon the irritability of the muscular tissue, and consequently over contraction, has been long known.

K

SECT. III. *On the Manifestation of Heat during Muscular Contraction.*

BECQUEREL and BRESCHET[a], in their ingenious experiments upon the relative temperature of the different tissues in the living animal, also found that, *during* muscular contraction, the temperature of the muscular tissue was increased. The amount of deflection was equivalent to about 1° of Fahrenheit.

MATTEUCCI[b] took two flasks, in each of which he suspended the limbs of five frogs, and in the middle of these limbs a delicate thermometer. Having waited until the temperature of the limbs had become stationary, he then caused the muscles in one of the flasks to contract violently. In the first experiment, the temperature of the air was + 12°60, the thermometer in the middle of the frogs indicated + 13°10; upon making the muscles contract for five or six minutes, the latter indicated + 13°50. At this point it continued stationary during eight or ten minutes. In another experiment, the temperature of the air was + 12°80, in the middle of the frogs + 13°50, during the contractions it ascended to + 13°95. An hour afterwards, the temperature being lowered to + 13°40, he then excited the muscles to contract, which they did although feebly, and the temperature only indicated + 13°60. In a third experiment, the temperature of the air being + 12°70, in the middle

[a] Becquerel, tom. iv. p. 21.
[b] Annales de Chimie et de Physique, Juin, 1856.

of the frogs + 13°55; after five minutes of contraction the temperature was increased to + 14°.
A current was passed fourteen hours after the experiment, but neither contractions nor any increase of
temperature were observed.

The results arrived at in the present chapter will
lead to the following conclusions.

First, *That during Muscular Contraction, Electricity
is evolved.*

Secondly, *That during Muscular Contraction, Chemical Action takes place.*

And, thirdly, *That during Muscular Contraction,
Heat is evolved.*

MATTEUCCI says, " Je pense qu'on doit désormais
admettre, comme prouvé par l'expérience, que l'action
chimique de la respiration musculaire pendant la
contraction est la cause de la force qui se développe
dans les muscles."

Now it does not appear to me that we would be
justified in considering one of these circumstances
either as a cause or as an effect of the other; that
the chemical action, for instance, should be the cause
of the electricity, or *vice versâ.* The muscular fibre
during life possesses an electric state, exists in
a state of tension, and during contraction this state
is lowered, and the force becomes free. This state
of tension is again restored by nutrition. Chemical
action takes place and heat is evolved at the same
time, and together with the evolution of the electric

force *during* muscular contraction. It would be a very interesting question to ascertain whether these three effects, the developement of electricity, of chemical action, and of heat during muscular contraction, took place according to any definite ratio. It would also, as observed by MATTEUCCI, be extremely important, "de s'assurer si l'activité de l'organe électrique de la torpille qui ne donne lieu qu'à une décharge électrique, sans chaleur et sans travail mécanique (contraction), est accompagnée ou non d'un phénomène analogue à la respiration musculaire. L'expérience est difficile mais important."

CHAP. XI.

Is nerve force a polar force? If so, in what respects does it differ from other polar forces? or what evidence have we that it is a polar force at all? These are questions which no doubt arise in the minds of many physiologists of the present day. Without entering into any detailed account of the various opinions that are entertained in regard to the nature of nerve force, it may be fairly considered, that those who entertain the opinion that nerve force is a polar force, are called upon, in the present state of the question, to give some direct experimental evidence in support of their conclusion: the *onus probandi* evidently rests upon their shoulders. We cannot remain satisfied with the vague notions that animal life and electricity are identical, or that nerve force and electric force are identical, or that nerve force differs entirely from electric force, or that vital forces are totally distinct from inorganic forces. The mind will not and cannot rest satisfied with these vague assertions; it requires something more definite. We want to know, if possible, how far they agree, and in what respects they differ. To answer these questions satisfactorily, it will be seen, that some of the most difficult physiological problems are now presented for our consideration; and although we might not be enabled to succeed in solving them to the extent we could wish, it is to be hoped that the attempt will be neither useless nor unprofitable.

The first question that will arise is the following: *Can we detect any manifestation of* CURRENT FORCE *in nerves* DURING *nerve action?*

It will be necessary to make a few preliminary observations respecting the employment of the terms *nerve action* and *nerve current*.

The fact that current force exists in a nerve is well known, and may be shewn, as was first pointed out by Du Bois Reymond, by placing the electrodes of a galvanometer, one in contact with the *tranverse* section of the nerve, and the other in contact with the *longitudinal* section or side of the nerve: this has been designated as the *nerve current*. So far this *nerve current* does not differ from that which has been called the *muscular current*, and which may be obtained in the muscle in the same manner. I have been led to consider, that this so-called *nerve current* is dependent upon *nutrition*. It does not afford direct evidence that the force which exists in the nerve as nerve force, and which may be supposed to be transmitted from one part of the system to another along the nerve, is *current force*: it only proves that the nerve tissue, like the muscular tissue, is in an electric condition—a condition no doubt necessary and essential for nerve action as well as for muscular action. Current force is not manifested when the electrodes are placed at the extremities of a nerve, the nerve being in a quiescent state; and this fact does not prove that the force transmitted by the nerve *during* nerve action is *not* current force. The real question therefore is, whether *during* nerve action, during the passage of nerve force along a nerve, current force is manifested or not. We must take care, however, and bear in mind, that the term *nerve current* may be employed to express two distinct ideas : 1*st*, The *electric current* manifested in the nerve, and which indicates the electric condition of tissue; and, 2*nd*, The *force—nerve force*—which is transmitted along a nerve *during* nerve action.

VAVASSEUR[a] and BERANDI[a] and DAVID[a] appear to have obtained some results in their experiments when the electrodes of a galvanometer were inserted into different parts of a nerve. These experiments, however, were undertaken prior to the knowledge of the existence of the nerve current, and the effects then observed may have been due to the electrodes having been placed on different parts of a nerve, the result being the so-called nerve current of DU BOIS REYMOND.

The experiments of PACINOTTI and PUCCINOTTI, repeated and confirmed by MATTEUCCI, must not be passed over. These inquirers inserted one electrode

[a] MULLER's Elements of Physiology; translated by Baly. Second Edition, vol. i. p. 685.

into the brain, and the other into the muscles, when
a deviation of the needle to a large amount occurred,
the electrode in contact with the muscle being
positive to the other. In this instance, it might be
supposed that the current travelled *from* the brain
along the nerve *to* the muscle. I have alluded to
these experiments in chap. vi.; and as similar effects
upon the needle were obtained when the electrode
was inserted into the internal jugular vein instead
of the muscle, I have been led to suppose that the
effects are due to the electric condition of the
nervous tissue, the result of nutrition, and not as
indicating the passage of an electric current along
the nerve.

PREVOST[b] and DUMAS[b], PERSON[b], MULLER[b], and
MATTEUCCI[c], have failed to obtain any evidence of
current force being manifested when the electrodes
of the galvanometer were inserted in the nerves of a
living animal. Some of my earliest experiments
were undertaken for the purpose of ascertaining this
point; and although results were occasionally obtained, nevertheless, as the experiments were performed prior to the knowledge of the existence of
the electric condition of the nervous tissue, I have
no doubt that the effects then observed were due to
the circumstance that the electrodes were placed on
heterogeneous parts of the nerve, as I shall be able
to shew presently.

[b] MULLER's Elements, pp. 686—689.
[c] Traité des Phénomènes Electro Physiologiques, p. 253.

My more recent experiments may be classed under three heads: 1st, Those in which the *galvanometer* was employed; 2nd, Those in which the *galvanoscopic frog* was used; 3rd, Those in which a *magnetized needle* was used.

Sect. I. *The Galvanometer.*

The animal, most frequently a rabbit, but occasionally a guinea-pig or a frog, was pithed, or rendered insensible by means of prussic acid. The sciatic nerve was carefully exposed throughout its whole course, and a plate of thin glass passed beneath it. All traces of blood being carefully removed, the pointed extremities of platinum electrodes were inserted at the extreme ends of the exposed nerve, as far apart as possible, leaving, however, a small portion, just at its exit from the pelvis, for the purpose of serving as a point of irritation. The leg of the limb was fastened down so as to prevent a too great motion during contraction, which might otherwise disturb the position of the electrodes. The other ends of the electrodes rested on and dipped into wooden cups containing mercury, the cups being placed upon glass for insulation: by these means a slight motion of the electrodes would not disturb the galvanometer, and it remained perfectly steady. The galvanometer consisted of several coils.

At first the nerve was irritated with the point of a steel needle, which was insulated, being inserted by means of a cork into a glass tube, which formed the handle; but, as may be readily supposed, the action of the steel needle upon the needles of the galvanometer so interfered with the results, that it was obliged to be set aside, and a pointed piece of copper wire, or a piece of glass or a glass pen, was used for the purpose.

The experiment thus arranged, the nerve was then irritated, and the muscles of the limb made to contract; but no effect occurred upon the needle. When, however, one electrode remained in the nerve, and the other was placed on its external surface, then the ordinary effect, the *nerve current*, was produced, the nerve during this period not being stimulated to action: if the nerve was stimulated and the muscles contracted, there did not appear to be any effect upon the needle indicative of an *increase* in the *nerve current*, or even of a *sudden decrease* in it, the needle gradually receding to its former position. The nerve was divided at its lower extremity, one electrode brought into contact with the divided surface, and the other with the longitudinal surface; the effects were the same; the *nerve current* appeared, but after that, there was no *increase* or indication of a current upon stimulating the nerve.

It will be readily seen, that two questions are involved in this last experiment: 1*st*, Is a nerve, during nerve action, traversed by a current of electricity? which we are now considering: and,

2ndly, Is the *nerve current* affected during nerve action ? a question which will be considered further on.

In other experiments the abdomen was laid open, so as to expose the lumbar plexus of nerves, and the nerves excited by a current from a pair of Grove's cells, so as to produce a more powerful contraction of the muscles ; the effects were negative, so long as the electrodes remained in their first position *in* the nerve ; but if, from the motion of the limb, or intentionally, the electrodes were moved, so as to be in contact with heterogeneous parts of the nerve, then vibrations of the needle were occasionally produced.   The spinal cord was irritated in the lower part of the dorsal region, by passing a copper wire between the vertebræ, so as to excite contraction, but the results were still the same. The nerve was excited by touching it with caustic potash, without any effect being produced upon the needle of the galvanometer; but if the alkali came into contact with one of the electrodes, then an effect occurred upon the needle, evidently due to the chemical action thus set up.

The animal was poisoned with strychnine, and as soon as tetanic contractions occurred, the experiment was repeated, but with the same negative results.

The only conclusion to be drawn from these experiments is the following : That *when the electrodes of a galvanometer are inserted* IN *a nerve during nerve action, there is no manifestation of current force ; but if*

*the electrodes come in contact with heterogeneous parts
of the nerve, during nerve action or not, then current
force is manifested; this effect being, however, the result
of the so-called* NERVE CURRENT.

SECT. II. *The Galvanoscopic Frog.*

The galvanoscopic frog is, unfortunately, accompanied with great uncertainty in its indications, and consequently requires the greatest care in its employment. The delicacy of its indications does not always correspond with its freshness; the muscles may be easily excited to contract when first prepared, but immediately afterwards it may be difficult, or even impossible, to arouse the contractions. This difficulty may arise from the muscles having once contracted remaining so. Again, a limb which is not very delicate in its indications at first, may, after some little time, be all at·once seized with slight tetanic contractions; this circumstance may arise from the nerve becoming dry, as has been already pointed out by MARSHALL HALL[d]. There are other circumstances which are undoubtedly influential, and amongst them the vital condition of the animal. It has frequently happened to me, that a limb which has been laid aside as worthless, has subsequently proved upon trial to be most delicate in its indications, without my being able to account for it. The muscles should be neither flabby nor too much contracted. The only test to be relied

[d] Edinburgh New Philosophical Journal, April, 1848.

upon is, by occasionally trying the limb with some well-known source of electric action, such as the muscle or nerve, and to have several at hand in case of emergency.

The former experiments were now repeated, and the nerve of the galvanoscopic frog, its limb being supported on a piece of glass, was laid at one time transversely on the sciatic nerve, at another time longitudinally on the nerve, and the nerve then irritated as before, but no effect occurred upon the galvanoscopic limb. If the galvanoscopic nerve be placed carelessly upon the other nerve, so that its end and side touch the nerve, and thus form a circuit, then the galvanoscopic limb will contract, this being the result of the nerve current in the galvanoscopic frog acting upon its own muscles; the effect being over, and the nerve remaining in this position, no contraction occurred upon stimulating the sciatic nerve; but upon *opening* the circuit of the nerve, it occasionally happened, if the frog was delicate in its indications, that contractions ensued. Should there be any blood upon the nerve of the animal, and the nerve of the galvanoscopic limb touch it and the nerve, then the galvanoscopic limb will contract; this may be considered as the result of a current arising from the contact of heterogeneous substances. But in neither instance was the effect any thing like that which occurs when the galvanoscopic nerve is placed upon a tetanized muscle.

Neither MULLER[e] nor MATTEUCCI[e] have been able to obtain any effect when they placed the galvanoscopic nerve upon another nerve in a living animal.

The sciatic nerve of the animal was now divided at its lower extremity, and the nerve of the galvanoscopic limb arranged thus : its divided extremity was placed in close contact with that of the animal, and then brought round so that its side should be in contact either with the side of the nerve of the rabbit, or with that of its own nerve, forming a loop. Now if there was any thing like a current passing along the excited nerve of the animal, we should expect it to be continued on along the nerve of the galvanoscopic frog, producing the effect of a *direct* current. The galvanoscopic limb contracted when first applied, due either to its own nerve current, or to that of the animal, but no contraction occurred during the stimulation of the nerve, unless the circuit was broken. Instead of the side of the nerve of the galvanoscopic frog being placed in contact either with its own nerve or that of the animal, the tendo Achillis was used : the effects however were still the same, no contraction during the stimulation of the nerve. When the end of the nerve of the galvanoscopic frog was merely placed in contact with the end of the sciatic nerve, so as to form a prolongation of the nerve, no contraction of the galvanoscopic limb occurred. The divided ends of the sciatic nerve of the animal were brought into as close

[e] *Loc. cit.*

contact as possible by bending the limb upon the thigh, and the nerve then stimulated at its upper part, but there was no contraction produced in the lower limb.

The same conclusion may be drawn, and the same remarks made, in regard to the galvanoscopic frog, as were made in regard to the galvanometer, viz. *when the proper precautions were taken, no evidence indicative of the manifestation of* CURRENT FORCE *in a nerve* DURING *nerve action could be obtained.*

SECT. III.   *The Magnetic Needle.*

It may be readily supposed, that if we failed to obtain any evidence of the existence of current force in a nerve during nerve action by means of the galvanometer, it would most probably happen that the magnetic needle would also fail to detect it. To exhaust every possible mode of its detection, the following experiments were undertaken. I may just add, that PREVOST[1], DUMAS[1], and MULLER[1], have already performed similar experiments, but failed in obtaining any result.

A small magnetised needle, three quarters of an inch in length, was suspended by means of a single fibre of silk-worm silk, the needle being passed through a strip of card, and to this the silk was attached. To avoid motion of the needle from slight draughts of air, and from vibration of the room, the silk was attached to a firm support, and a glass tube, an inch in diameter and five inches in length, was

[1] *Loc. cit.*

so arranged as to inclose the needle. The sciatic
nerve being laid bare, a plate of glass was placed
beneath it, and thus the nerve was elevated above
the surrounding muscles; it was then brought
beneath the magnetic needle, and kept in that
position, being supported upon a stool about a
quarter of an inch below it. The nerve was placed
in various positions with regard to the needle, sometimes parallel, at other times transversely or ob-
liquely to it. The needle being perfectly steady, the
muscles of the leg were made to contract, as in the
previous experiments; but, in whatever manner the
experiment was arranged, there was no indication of
any action upon the needle. Whenever any motion
of the needle occurred, it was evidently due to the
motion of the atmosphere produced by a too great
motion of the limb. I may also observe, that, as the
experiment was performed near a window, it was
necessary to guard against the heating effects of the
sun's rays upon the air within the tube. To the
importance of attending to this latter circumstance
in these experiments I have already alluded.

Can we magnetise a needle? Thick and thin needles
about three quarters of an inch long, and free from
magnetism, were inserted either *transversely* or *longitudinally* in the nerve, and the muscles then made to
contract powerfully, and for some time by means of
an electric current from two of Grove's cells. In
these experiments there was no distinct evidence of

the needles having become magnetised. Upon testing the needles after the experiment with steel filings, it was frequently observed that the filings adhered to the needles throughout its whole extent, although they had been previously wiped with a dry clean cloth. These effects were evidently due to the needle being damp; the moisture from the fingers was sufficient to produce this damp state, and the only mode of preventing it was by heating the needle, but by so doing, any magnetic state it might possess would be destroyed. The difference observed in the needles in which the filings adhered when magnetised or when damp is very great; in the latter the whole surface is slightly covered with them, but in the former the filings are confined to one or two spots.

The results of the present investigation only tend to confirm the opinion already expressed by MULLER, MATTEUCCI, and others, that *nerve force* is not identical with *current force*. Or the conclusion may perhaps be more correctly expressed by saying, that *we have not been enabled to obtain any evidence of the manifestation of current force in a nerve during nerve action*. It must be borne in mind, however, that I am now speaking of nerve action, and not disproving the existence of the so-called *nerve current* which is manifested in a nerve during its quiescent state: and this brings before us another most important question for consideration, viz. *Is this nerve current affected during nerve action?*

L

In DE LA RIVE's valuable Treatise on Electricity[g], and to which I must refer, as containing perhaps the most recent views in regard to this question, will be found the following important remarks: "We know," says DE LA RIVE, " that the nerve possesses of itself a certain electrical state, which we have succeeded in determining; we know, moreover, that this electric state is modified by every excitation exercised upon the nerve..... Now if by any cause whatever the electric state of the nerve is modified, equilibrium is destroyed; and from this there results a contraction of the muscle, or a sensation. Before studying the consequences of the modification, we may remark, that it consists in the fact that the organic molecules of which the nerve is formed are not polarised transversely from within outwards, but longitudinally from one extremity to the other, as is every conducting body traversed by an electric current. When the modification arises from the immediate action of the nervous centre, it appears that the polarisation is brought about always in such a manner, that the negative poles of the molecules are turned on the side of this centre, and the positive on the side of the muscle, as would result from the action of an electric current that might be travelling in the direction of the nervous ramifications. This it is that explains why an electric current which travels in this direction favours the contraction much more

[g] A Treatise on Electricity, by Aug. DE LA RIVE. Translated by C. V. WALKER, F.R.S. vol. iii. p. 56.

than when it travels in the contrary direction. This
is equally a natural consequence of the fact, that the
particles of the nerves upon which the immediate
action of the brain must be exerted, being the
interior which penetrate into it more deeply, have
their negative poles free.

" If, instead of coming from the brain, the action
exerted upon the nerve comes from the muscle, the
polarisation of the nerve must take place in a
contrary direction, namely, so that the positive poles
are all turned towards the side of the nervous
centre, and the negative towards the side of the
muscle whence the excitation comes." I have
quoted these observations at some length, as containing perhaps the most recent views upon the
subject, and supported by an authority of some
weight. Now the only conclusion that appears to
me that can be drawn from these remarks is this,
that the force transmitted along a nerve during nerve
action is identical with current force as it exists in a
wire carrying a current of electricity. If so, I need
scarcely add, that we ought to be enabled to obtain
some direct *experimental* evidence in support of this
opinion : at the present time I know of none beyond
that advanced by Du Bois Reymond, who appears to
have ascertained in his experiments, that the *nerve
current* may be made to *increase* or *diminish* according
as the nerve is excited by an electric current—being
increased if the current passes in one direction, and
diminished if it passes in the contrary direction.

I have already had occasion to relate some experiments in chap. ix. in which I endeavoured to obtain an *increase* of the nerve current according to the mode suggested by Du Bois Reymond, but entirely failed in doing so. As they were not undertaken for the express purpose of ascertaining whether the nerve current is affected during nerve action, the following were performed.

Is the nerve current affected during nerve action? Instead of employing a nerve separated from the animal, the experiments were conducted in a manner similar to those that have been already related in the previous part of the present chapter, viz. with the sciatic nerve. The electrodes of platinum were coated at one extremity with shell-lac, leaving, however, the extreme end bare, and one of them was pointed so as to be easily inserted into the substance of the nerve, whilst the other electrode presented a flat surface to rest upon the surface of the nerve. The distance at which the electrodes were placed from each other varied from an inch to a quarter of an inch, but were generally within about half an inch of each other. When the needle indicated the existence of the nerve current, the upper end of the sciatic nerve was stimulated either by means of the glass pen, or copper wire, or an electric current, to produce muscular contraction in the leg: sometimes the nerve was stimulated by a constant current, at other times by an intermitting current. The current was passed at one time as a *direct* current, at another time in the *inverse* direction. Now in whatever manner the nerve was excited to action, I failed to obtain any evidence of a decided *increase* in the *nerve current*, neither could I obtain any definite indication of a *sudden decrease* in the nerve current; the needle gradually receded. Vibrations of the needle were frequently observed, and were evidently due to the motion of the electrodes caused by the movement of the limb during the contraction of the muscles. The effects upon the needle were just the same, in whatever position the electrodes of the galvanometer were placed ; whether the electrode in contact with the surface of the nerve was placed on the upper or lower portion of the nerve, between the stimulated portion of the nerve and the other electrode, or below the latter. There was no decisive action upon the needle in these experiments indicative of any marked influence over the nerve current; the needle merely returned to its former position, or gradually receded.

When the galvanoscopic frog was employed, as in Sect. II., after the first effect of the nerve current was over, there was no further contraction, however long the nerve was stimulated; there was no effect corresponding to the tetanized muscle; it was impossible to produce a tetanic condition of the nerve, so that it should affect the nerve of a galvanoscopic frog [h].

[h] What the condition of the nerve may be along which the *inverse* current has passed for some time, so as to produce

I should certainly hesitate before coming to the conclusion, that *no* effect is produced upon the nerve current during nerve action; but I certainly have not been able to obtain those definite and constant indications that we have a right to expect, under the supposition that the nerve current, as manifested in the *transverse* direction of the nerve, is converted *during* nerve action into a current in the *longitudinal* direction of the nerve, which appears to be the opinion of DE LA RIVE, as expressed in the remarks I have already quoted.

Concluding Remarks.

The conclusions that we have arrived at in the present investigation being of a negative character, may perhaps be considered as any thing but satisfactory. Let us not, however, be led away by the false supposition, that because negative results have only been obtained, that therefore no positive knowledge is acquired: we may have ascertained a most important fact, if true, and, whether true or not, may partly depend upon our being able to give a satisfactory reason for our failures. Have we not, it may be asked, commenced our inquiry with a false notion in regard to nerve action; viz. its identity with electric action? Have we not supposed an identity to exist between *current force* and *nerve force*, which we have failed to prove?

tetanic contractions in its own limb, I have not been able to ascertain; there is no *increase* in the nerve current.

To suppose that the conditions may exist in the
one case and not in the other, and that the two
forces may still be the same, cannot be deemed
satisfactory to any experimentalist. If nerve force
be electric force, we have a right to ask for some
proof of it. The circuit form of the arrangement
exists in the *transverse* direction of a nerve, and we
can detect the necessary current, the so-called *nerve
current.* The effects here, however, are analogous
to those of a charged Leyden jar rather than to a
voltaic circle, as I have endeavoured to shew in
previous chapters, and to which I have already
referred. The very circumstance of our being able
to shew this state in the *transverse* direction, would
lead us to expect that we ought to be able to prove
its existence in the longitudinal direction, if it
existed.

The subject under consideration involves, however, three distinct questions. 1*st,* Is nerve force
nothing more than the electric force which exists in
the nerve, and put into motion *during* nerve action?
or, 2*ndly,* Is this electric force in the nerve *converted,*
as it is called, into nerve force during nerve action?
or, 3*rdly,* Is this electric condition of the tissue
merely a condition, and perhaps a necessary condition, for the manifestation of nerve action?

In reply to the first question, it may be observed,
that my experiments have failed to give the necessary
proof in favour of this supposition.

In regard to the second question, if we could

obtain any decided evidence of the *nerve current* being affected *during* nerve action—I say decided—then we should be called upon to account for its *increase* or *decrease*, according to the principles of *conservation of force*[i], and to shew what has become of the force. But if the effects that have been obtained with the galvanometer, viz. a *gradual decrease* in the nerve current, such as I have observed, be due merely to a disorganization of the nervous tissue, and they are such as are consonant with this supposition, then nerve force must be ranked as a higher *form* of force, and the electric condition of the tissue merely a condition for the manifestation of nerve action. To assist in elucidating this question, let us just refer for a moment to the muscular current, and see whether this current is affected during muscular contraction.

When the electrodes of a galvanometer are so arranged with the muscular fibre, viz. in contact with the transverse and longitudinal sections, the muscular current is produced ; upon making the muscle contract, the needle returns rapidly to its former position, and passes beyond it. Du Bois Reymond has designated this effect, " the negative . variation of the muscular current;" the exact meaning of the phrase I do not comprehend, if it be intended to express more than the fact. The question, however, is this, Is the return of the

---

[i] On the Conservation of Force. By Prof. Faraday. " Philosophical Magazine," April, 1857.

needle in consequence of one of the surfaces of the
muscle separating from the electrode during the
contraction of the fibre, and so breaking the circuit;
or is there a sudden diminution in the electric
tension of the tissue, and so producing a *decrease*
in the muscular current? Now we have some
evidence that there is a discharge of electric force
during muscular contraction, as shewn by the galvanoscopic frog in MATTEUCCI's experiments; but we
have no evidence that the muscular current is increased in these experiments during muscular contraction. The loss of force, the lowering of the
electric tension of the muscular tissue, which takes
places during contraction, is restored by nutrition[k].
If we can detect the evolution of electric force
in the muscular tissue during muscular action,
surely we ought to be able to obtain some evidence

[k] According to the principles of the *Conservation of Force*, we
must endeavour to trace out in what manner the force in the
muscle is disposed of. We have *heat* developed during muscular
contraction, as shewn by BECQUEREL, and BRESCHET, and
MATTEUCCI. Carbonic acid is evolved during muscular contraction, as shewn by MATTEUCCI in his experiments on muscular respiration. Electric force is evolved during muscular
contraction, as has been proved by MATTEUCCI, and confirmed
by others, and also by some of my own experiments. It would
be, perhaps, erroneous to suppose, that the force is converted
into chemical action, as nutrition; nutrition is undoubtedly
increased during continued muscular exertion; but muscular
action—contraction—would appear to be the first act in this
series of events, and nutrition the second; contraction may be
considered as the exhausting act, nutrition the restoring act,
in this process.

of the manifestation of electric action in the nervous tissue during nerve action, if nerve action be, as it is supposed to be, merely the result of the electric force of the tissue being converted into current force.

Let us not hastily conclude that nerve force is totally distinct from electric force, or that it bears no relation to it or the other *polar* forces, such as magnetism, for example. We appear to be in a position somewhat similar to the physical philosophers prior to the discovery of FARADAY of magneto-electricity. So long as they confined their experiments to the mere application of the electrodes of the galvanometer to the two ends of the magnet, no result was obtained; the necessary evidence, the connecting link, was wanting; and we may now be in a somewhat similar position. I am not now supposing, or going so far as to say, that it *must* be by means of the galvanometer that this necessary evidence, the connecting link, is to be obtained; it *may* be by some other fact totally unconnected with the galvanometer that the connection will be shewn. We have sufficient evidence, however, to shew, that an intimate *connection* exists between nerve force and electric force, viz. in the development of the electric force in the fish, in the dependence of the former upon the *will* of the latter. We may also refer to the relation that exists between nerve force and muscular force. We have strong evidence for believing that muscular force is a *polar* force; the

electric condition of the tissue and the development
of the force during contraction is in favour of the
supposition. We are not compelled to assume that
any peculiar force exists, or is associated with the
muscle distinct from this electric force; only that the
mode in which it exists is brought about by other
agencies than those that occur in the inorganic
kingdom, viz. by *nutrition*. It is in these intimate
relations and connections in the dependence of the
developement of electric force in the fish, and of
muscular contraction upon nerve action, that the
strongest evidence for the *polar* character of nerve
force is manifested, and not so much upon the
electric condition of its tissue[1].

But, it may be asked, are we justified in supposing that nerve force, in nerve action, is entirely
independent of the electric force as it exists in the
nerve in its quiescent state? Do we not see, that
nerve force bears some relation to the state of the
vital powers of the animal, viz. its nutrition? as
nutrition is increased or diminished, does not
nervous energy increase and diminish in a corresponding ratio? and have we not *experimental*
evidence to shew, that the electric state of the tissue
(*the nerve current*) is dependent upon, and varies

[1] We must not overlook the connection which exists between
nerve force and *secretion*. The latter is undoubtedly a *polar*
action, but the influence of nervous action over secretion,
although it undoubtedly exists, does not appear to be exerted in
so immediate and *direct* a manner as in the two instances I
have just quoted.

according to, the state of nutrition of the nerve tissue? Do not all these facts shew, that an intimate relation exists between these two forces—between the electric force in the tissue and the nerve force—and therefore what right have we to suppose that any other force but the electric force is connected with the nerve?

All this may be granted, but the premises do not justify the conclusion. There may be no necessary connection between the electric state of the tissue and nerve force beyond that of being a condition, and perhaps a necessary condition, for the manifestation of nerve force. I will not go so far as this, and say that the electric force of the nerve has *no* connection whatever with nerve force; but only this, that nerve force is *not* merely the electric force of the tissue converted into current force, as it has been supposed. Whether the electric force of the tissue is *converted* into nerve force *during* nerve action is another question, and perhaps the question for us to solve. At present, we have no *experimental* evidence to prove this supposition; for we have not obtained any indication of a decided loss in the electric state of the tissue, a *sudden decrease* in the nerve current *during* nerve action, to indicate such a *conversion*. The evidence, however, is not sufficiently decisive of the question, and at present it may be considered an open one[m].

[m] I have not thought it necessary to adduce other well-known evidence to shew the difference between the action of nerve

It must be remembered, that we have hitherto limited our views, and have been comparing nerve force with only one of the ordinary polar forces with current force. Let us now compare it in regard to magnetic action. In a magnet we can get no evidence of current force travelling *from* one pole *to* the other; the force exists in a state of tension, which may be raised or lowered without affecting the galvanometer when arranged with the electrodes at the two ends. May not nerve force exist in a state of tension, and nerve action correspond with a rising or lowering of this state, just as the muscular fibre may be regarded as existing in a state of electric tension and muscular contraction, the result of a lowering of its tension? The developement of the force in the fish, and other facts, may perhaps be adduced as an argument against this supposition, and in favour of transmission of force *from* one point *to* another along the nerve. But even here an increase or a lowering in the tension of one part may be accompanied with a lowering or an increase in the tension of another part; the final result being tantamount to the *transmission* of force from *one* part *to* another. Still it behoves us not to keep limiting our views to one class of actions only, but to be bold and suggestive, and seize upon any resemblances, however slight; and provided we do

force and of current force. The ligature of a nerve preventing the transmission of nerve force, is a strong argument against the identity of these two forces.

not allow them to take a too firm hold of our minds, so as to lead us to consider them as realities, but merely suggestive, by leading to future experiments, we may hope to arrive ultimately at some more promising and positive results.

The following conclusions may be deduced from the present inquiry.

First, That nerve force, *during* nerve action, is not *current* force.

Secondly, That the electric condition of the nerve, as manifested by the nerve current, is not *converted* during nerve action into currrent force.

Thirdly, That the electric condition of the nerve may be merely a condition, and perhaps a necessary condition, for the manifestation of nerve action.

Fourthly, That the evidence in favour of nerve force being *polar,* is shewn by the *connection* that exists between the developement of the electric force in the fish and its dependence upon the will of the animal, and also in the connection between nerve action and muscular action, the latter being regarded as *polar.*

Fifthly, That the whole of the evidence indicates that nerve force is of a higher character than any of the other known *forms* of *polar* force.   And,

Sixthly, The question, whether the electric force, as it exists in the nerve, may not be *converted* into nerve force *during* nerve action, may be considered at present an open question.

CHAP. XII.

THAT electric currents may be obtained by inserting the platinum electrodes of a galvanometer into
different parts of vegetables, has been proved by
BECQUEREL [a], DONNE [b], WARTMANN [c], ZANTEDESOHI [d],
and BUFF [e]. According to BECQUEREL [f], " The dis-
tribution of the ascending sap, and the liquid of the
cortical parenchyma, leads to the belief that currents
continually circulate in vegetables, from the bark to
the pith." . . . " The leaves act like the green part of

[a] Traité de l'Electricité, tome iv. p. 164; Comptes Rendus,
Nov. 4, 1850, Mai 5, 1851; or Philosophical Magazine, 1851;
Mémoires de l'Académie des Sciences de Paris, tome xxiii.
p. 301, 1853. I may refer also to a valuable Paper by Professor
GOODSIR, in the Edinburgh New Phil. Journal, Oct. 1855.

[b] Traité de l'Electricité, tome iv. p. 164.

[c] Bibliothéque Universelle de Genève, Dec. 1850; or Philosophical Magazine, 1851.

[d] In the Comptes Rendus for Mars 24, 1851, ZANTEDESCHI
refers to the Comptes Rendus des Sciences de l'Institut Vénitien
des Sciences, Lettres, et Beaux Arts, 26 Mai, 1850, as containing the results that he has arrived at. I have been unable to
meet with this publication.

[e] Philosophical Magazine for Feb. 1854, and Annales de
Chimie et de Physique, 3me Serie, tome xli. p. 198.

[f] Philosophical Magazine, 1851, p. 578.

the parenchyma of the bark; that is to say, the sap which circulates in their tissues is negative with relation to the wood, to the pith, and to the earth, and positive with regard to the cambium."....

" The chemical actions are the first causes, it cannot be doubted, of the electric effects observed in vegetables."

" In the roots," says WARTMANN[g], " the stems, the branches, the petioles, and the peduncles, there exists a central descending current, and a peripherical ascending current; I call them *axial* currents."....

" In most leaves the current proceeds from the lamina to the nerves, as well as to the central parts of the petiole and the stalk. In certain fleshy plants, it is directed from the medullary or cortical portions of the stalk towards the mesophyllum, and from the latter towards the superior and inferior surfaces."...." They arise from an electro-chemical action between the liquid substances brought into contact by the tearing of the tissues. The weak residual current (which is the normal current) owes its origin to the interposition of the porous vegetable walls between juices of different concentration, and proceeds through them from the densest to the least dense liquid."

The general conclusion that PROFESSOR BUFF appears to have arrived at is the following: " *The roots, and all the internal portions of the plant filled with sap, are in a permanently negative condition; while the moist*

g Philosophical Magazine, 1851.

*or moistened surface of the fresh branches, leaves, flowers,
and fruits, are permanently positively electric."* . . . .
He considers that " the electro-motive action arises
from the moist surface of the plant on the one hand,
and the liquids which are in its interior on the
other."

The results of these inquirers would lead us to
suppose, that the effect upon the needle is due to
what may be termed *secondary actions,* viz. to the
reaction of the different vegetable juices upon each
other, and to the reaction of the fluids upon the
surface of the platinum electrodes. The very fact,
however, that a difference in the fluids exists, proves
also that a force capable of causing this difference
must likewise exist; and the question naturally
arises, are not these *primary actions* accompanied
with the developement of electrical actions?

The *primary actions* I refer to are those of *secretion,
nutrition,* and *absorption.* Now as the leaves and the
roots perform some of the most important functions
in plants, it appeared probable that it would be in
these organs that we might obtain a solution of our
problem; and the two following questions now occurred: 1*st,* What would be the effect if the *external*
surface of the leaf, and the *sap* flowing from it, be
formed into a circuit? and, 2*dly,* What would be the
effect if the *external* surface of the root (the spongioles) and the *fluid* ascending from it be formed
into a circuit?

M

SECT. I. *On the Manifestation of Electric Currents in the Leaves of Plants during Vegetation.*

The mode of conducting the experiments was as follows: The leaf was placed upon a clean piece of glass, and the extremity of one platinum electrode, to the extent of half an inch or an inch, was placed upon the upper or under surface of the leaf; a small notch was cut in the petiole, and, as the sap flowed out, the extremity of the other electrode was placed in contact with it.

I may just remark, that after having worked for some time upon plants growing in pots and in London, I was led, from the unsatisfactory results that were obtained, to repeat the experiments upon plants growing in the open air and in the country; and it soon became evident that, in order to obtain any thing like satisfactory results, strong, healthy, and vigorous plants should be employed[h].

Experiment 1. Vegetable marrow. A healthy middle-sized leaf, and the sap from the petiole; the latter *positive*[i] 3°. A large leaf slightly tinged with yellow, and dry, and the sap from the petiole—no effect. Various leaves were tried with similar results; when any effect was obtained, the sap was *positive.*

[h] The experiments were performed during the months of July, August, and September, 1852.

[i] The difficulty experienced in comprehending the results obtained by different inquirers in reference to the *direction* of the current I have already alluded to in chap. x.

The surface of the leaf was occasionally moistened with water, with doubtful results as to the effect being increased. If the leaf had been separated from the plant for any time before the circuit was formed, no effect occurred.

As several of the experiments were carried on in the open air, one or two circumstances occurred which it was necessary to guard against. If the weather was at all boisterous, it became utterly impossible to continue the experiments, the slightest breeze being sufficient to shake the instrument. To obviate this difficulty, the galvanometer was firmly fixed upon a heavy block of wood, and sheltered. I was, however, perplexed by another circumstance. Working during a fine and calm day, it occasionally happened, that, just upon the point of completing a circuit, the needle would move to the extent of 10° or 15°, or more, without any apparent cause, the needle up to that time having been perfectly steady, and the circuit not yet completed. After some time it was noticed to occur just after a cloud had passed over the sun, and it was thought to be due to a slight breeze that might be then produced; but I am now disposed to consider it as owing to the heating effect of the sun's rays upon the glass shade of the instrument creating a motion of the air within the shade. Some facts bearing upon this question have been noticed by PROF. TYNDALL[k] in reference to some experiments of DR. GOODMAN.

[k] Philosophical Magazine, Feb. 1852.

Another fact worthy of notice may be mentioned. As the galvanometer was frequently obliged to be moved into the neighbourhood of the plant, it would sometimes happen, from the altered position of the instrument, that the plane of the coil would be in the reverse direction of that of the needles to which it had been previously; hence contradictory results would very readily be supposed to be obtained.

To avoid unnecessary repetition, I shall generalize the results that were obtained in the following experiments. They were conducted in the same manner as in the first experiment, and upon leaves of different varieties of the following plants: cucumber, vine, lettuce, cabbage, nasturtium, convolvulus, rose, ivy, hop, walnut, geranium, fuchsia, strawberry, bean, apple, ficus elastica, ficus carica, lemon, orange, oleander, eutaxia, camellia, mesembryanthemum, lily, marvel of Peru, pirus japonica, tropæolum, wistaria, elder, sycamore, hollyhock, arum, hydrangea, thistle, and dahlia. In some the effects were null, in others but slight; the greatest effect obtained amounted to about 2° or 3°; the electrode in contact with the sap, *positive*. The most satisfactory results appeared with the firm, compact leaves, such as the camellia, vine, elder, and sycamore, and when the electrode was in contact with the under surface. It frequently happened, that, in experimenting with different leaves of the same plant, effects might be obtained with one leaf but not with another, although in other respects perfectly similar[1]. This circumstance may be adduced as shewing, that the effect upon the needle cannot be entirely referred to the changes which occur in the sap from exposure to the air; and the question now arises, To what actions can we refer these effects? Is the sap *acid*, or does the sap contain a *cation?*

I shall not enter into any discussion as to the functions of leaves, whether they may be considered as organs of respiration, digestion, or absorption, but refer my readers to works on physiological botany. That oxygen gas or carbonic acid gas is given off by the leaves of plants, may be considered as having been experimentally proved.

If we break off the petiole of a leaf, and just touch a piece of litmus paper[m] with the divided surface, in

---

[1] It is probable that the circumstance of not obtaining any result upon some of the leaves, might depend upon their not being always in the same state of action. Some experiments on *the Respiration of the Leaves of Plants*, by MR. PEPYS, bearing upon this question, will be found related in the Phil. Trans. for 1843, p. 329.

[m] In the employment of litmus as a test we must bear in mind, that a compound might shew an acid reaction, without possessing at the same time an excess of acid. According to Berzelius, "the colour of litmus is naturally red, and it is only rendered blue by the colouring matter combining with an alkali. If an acid be added to the blue compound, the colouring matter is deprived of its alkali, and thus, being set free, resumes its red tint. Now on bringing litmus paper in contact with a salt, the acid and base of which have a weak attraction for each other, it is possible that the alkali contained in the litmus paper may have a stronger affinity for the acid of the salt, than the base has with which it was combined; and in

the majority of instances the sap will be found to have an acid reaction; in some succulent plants the effect will be null or very trifling, whilst in other plants, such as the vine, it will be very apparent. In the latter instances, the effects upon the needle may be fairly referred to the *acid* reactions of the secretions of the plant; but can we refer them to the same action when the sap does not present an acid reaction? Are we sure that the sap is acid? May not this acid reaction upon the litmus arise from the acid *secretions* of the plants, which become mixed with the sap upon the tearing asunder of the tissue of the plant?

The difficulty which arises in solving some of these questions occurs from the fact, that there is not that distinct circulation in plants as in animals; the sap, as it passes from one spot to another, may undergo a series of progressive changes, and in the tearing and cutting of the tissues, a mixture of sap from different parts and of secreted products must naturally occur. We cannot place our electrodes at the real *acting point;* there may be several acting points between the electrodes; and the resulting effect upon the needle may be that of a *differential* or of a *combined* current, according to circumstances.

Two solutions, one *acid*, consisting of 8 drops of strong sulphuric acid to one ounce of water, the

that case the alkali of the litmus being neutralized, its red colour will necessarily be restored." *Turner's Chemistry*, p. 671. 5th Edit.

other *alkaline*, consisting of 120 drops of the liq. potassæ (*Phar. Lond.*) to one ounce of water, were prepared, and used in the following manner: The electrode to be placed in contact with the sap was dipped into a portion of the *alkaline* solution, and then applied to the cut petiole; this electrode was still *positive* to the other, the effect, however, was not increased; if the effects were due in the former experiments to the acid reactions of the sap, we should now have expected the electrode to be *negative*. The experiment was repeated upon another leaf, but with the *acid* solution; the effect was now very much increased, and the electrode in contact with the sap *positive*.   Several other experiments of the same nature were performed, the results of which I shall generalize by stating, that whenever the electrode was coated with the *alkaline* solution, whether it was that in contact with the surface of the leaf, or that in contact with the sap, with but few exceptions there was no decided difference as to the effect upon the needle; but that whenever one of the electrodes was dipped into the *acid* solution, this was always *positive* to the other.

If the leaf was gathered, an electrode being inserted into the petiole, and the leaf then plunged into a glass of water containing the other electrode, it frequently happened that the electrode in contact with the water was *positive* to the other, even if the water was rendered alkaline.   It appeared to occur principally with the leaf of the bean (Windsor), and the strong-scented leaves, such as the geranium, &c.

In judging of the results obtained by means of these solutions [n], we must consider their effect under two points of view; 1st, as producing their own chemical effect; and, 2ndly, as forming a better conducting medium for the current. When the *acid* solution was employed, there can be no doubt that the current then obtained was due to the immediate action of the acid upon the tissue and juices of the plant; it always indicated its *positive* condition, and with increased effect; but how shall we account for the action of the *alkali?* This should have indicated a *negative* condition, if the current was due to its own immediate action; it was generally found, however, to indicate a *positive* state when in contact with the cut surface of the petiole; and here I cannot help believing but that it must have acted, under these circumstances, as a *conducting* liquid.

From these experiments we may deduce the following conclusions:

1st, That when the electrodes of a galvanometer are brought into contact, one with the surface of a leaf, and the other with the sap flowing from the same leaf, an effect occurs upon the needle, indicating the surface of the leaf and the sap to be in opposite electric states.

---

[n] I need scarcely point out the importance of paying particular attention in the use of towels, and to cleanliness, especially of the fingers and hands, when employing these solutions.

2ndly, That these effects cannot be referred entirely to ordinary electro-chemical actions; but that,

3rdly, They may be referred in part to the organic changes which take place in the leaf during vegetation.

Before quitting the subject in reference to the leaves, I shall make one or two observations. I have generalized the results of my experiments, and therefore, in their repetition, the same precise result must not always be expected. We must bear in mind, that in these experiments we are obliged to have a fresh subject for every experiment, and it is almost impossible to meet with that identity of circumstances we could wish; here the physical philosopher possesses advantages which are denied the physiologist. The problem we are endeavouring to solve is connected with the *vital actions* of the part; these terminated, nothing but difficulties then arise.

Sect. II. *On the Manifestation of Electric Currents in the Roots of Plants during Vegetation.*

Becquerel[o] states, that "currents exist going from the pith and the wood to the bark, by the mediation of the roots."

According to Wartmann[p], "when the soil and any part of a plant, visible or underground, is placed in the circuit of the rheometer, we find a current directed

---

o Philosophical Magazine, 1851.
p Ibid.

from the plant to the soil, which is thus positive with relation to it." . . . . . " The superficial layers of the soil are frequently positive relatively to those which surround the spongioles."

In several experiments, the facts observed by BECQUEREL and WARTMANN were obtained, viz. the electrode in contact with the soil was *positive* to that in contact with the plant; but that the effect was in a great measure due to the soil, was shewn thus,— both electrodes were inserted into the soil, when one was found generally to be *positive* to the other. To obviate this difficulty, the electrode employed to be inserted into the earth was coated with sealing wax to the extent of three inches, leaving the extremity exposed for about a quarter of an inch.

Several experiments were performed upon plants, such as different kinds of geraniums, fuchsias, balsams, antirrhinums, lophospermums, and vegetable marrows, growing in pots, in the following manner: The coated electrode was inserted into the soil, and a notch having been cut in the stem about an inch above the earth, or the stem entirely removed, the other electrode was then brought into contact with the divided surface; sometimes the electrode was inserted into the stem; the general result was as follows: in a very few instances, with a lophospermum and an antirrhinum, and one or two geraniums, the electrode in contact with the trunk of the plant, with the sap, was found to be *positive* to the other; in many instances no decisive effect was obtained; but in the majority of instances the electrode in contact with the soil was *positive* to the other. The influence of the soil was shewn thus: in those plants which had been recently repotted, and the mould consisted of a mixture of sand and peat earth, the effects were null, and the electrodes when inserted alone in the pots indicated but slight effects; whereas in those pots in which the earth had remained some time, and consisted of common garden mould, the effects were very decided. Dampness of the soil also increased the effect.

As it was difficult to know when the electrode was in contact with the spongioles, the following method was adopted.

Broad beans (Windsor) were made to vegetate in the dark in water, and in small glass jars containing but a small quantity of mould. At different periods of vegetation, the apex of the plumula or the stem was removed close to the cotyledons, and a circuit formed between the divided surface and the external surface of the radicle; the latter was generally *positive;* in some instances no effect ensued. Supposing that the effects might arise from the acidity of the soil, a. dilute solution of potash was poured into the glass jar. Under these circumstances, however, the electrode in contact with the soil was *positive* to the other. The whole root was taken out of the soil, and washed gently in water so as to remove the earth, then placed in water, and the circuit reformed; the electrode in contact with the water was still *positive.* According as the surface of the root in the water was greater, so was the effect upon the needle increased. The sap from the roots reddened litmus paper.

Hyacinth and narcissus bulbs were made to vegetate in the usual glass vessels. The electrode in contact with the water and at the extremities of the roots was *positive* to the other when inserted into the stem or bulb. The bulb was raised so as to leave the fibrillæ alone in the water; the effect was much diminished, and almost null. The water was rendered slightly alkaline by means of potash; the effects were still the same.

The sap in the fibrillæ and bulb reddened litmus paper.

The same experiments were made as with the leaves, namely, coating the electrodes with acid or alkaline solutions. Whichever electrode was coated with the acid, was always positive to the other.

BECQUEREL[q] has made the following remark: " On obtient peu ou point d'effet, lorsque l'une des aiguilles est dans le ligneux, près de la moëlle, et l'autre dans la terre."

According to LINDLEY[r], Brugmans has ascertained that some plants exude an acid fluid from their spongioles.

Some experiments are also related by BECQUEREL[s],

---

[q] In his first Memoir.
[r] Introduction to Botany, p. 229.
[s] Traité de l'Electricité, tome iv. p. 185.

in which it would appear that an acid is formed during the germination of the seeds.

If any difficulty occurred in pointing out the *electro-positive element* in the leaves, the same difficulty meets us here, and the question arises, Are not the results which were obtained with the galvanometer rather the results of *secondary reactions*, in the majority of instances, than those of the *primary actions* (the normal results), and which I obtained but in a few instances, viz. when the sap indicated a *positive* condition? The very circumstance of an acid being secreted or formed at the root would indicate that an *electro-positive* element must have appeared somewhere, and why not, it may be asked, might it not have been absorbed by the plant? Might not those instances in which I obtained no effect upon the needle arise from the *two* currents being so equally balanced, that the resulting effect upon the needle was null? In judging of the effect obtained by the needle‡, it appears to me that very erroneous conclusions would be arrived at, if the greatest amount of effect were to decide in each case; we must take into consideration the assumed *origin* of the current, and see whether the effect bears any proportion to the amount of force in

‡ There is also something to be observed in the *motion* of the needle, which can only be obtained by practice, when judging of slight effects. Far more correct results are indicated when the needle moves steadily and in a constant and definite manner, than when a sudden and great amount of motion is obtained.

action, and also those circumstances which are likely to influence the normal result, before drawing our final conclusions.

From these results we may deduce the following conclusions :

1st, That when the electrodes of a galvanometer are brought into contact, one with the external surface of the spongioles of a plant, and the other with the sap ascending from the roots, the sap and the external surface are in opposite electric states: and, 2dly, That the effects which are observed with the galvanometer may, in the majority of instances, be due to ordinary electro-chemical actions, but that, in some instances, the effect cannot be referred to these actions, but may be referred to the organic changes which occur in the fluids in the roots during vegetation.

SECT. III. *On the Manifestation of Electric Currents in the Petals of Flowers during Vegetation.*

According to WARTMANN[a], " the currents are feeble in flowers."

Circuits were formed between the surfaces of the petals and the sap from the peduncles, in the following plants : Geraniums, *various;* nasturtium ; balsams, single and double; fuchsia; hollyhock; convolvulus; vegetable marrow; and cucumber. The effects obtained were but slight, and in many instances null. Whenever the effects were obtained,

[a] *Loc. cit.*

they indicated the sap to be *positive*. The greatest effect appeared with the fuchsia.

It may be a question whether the results were not due more to the changes which the sap might undergo from exposure to the air, than to the changes which might occur in the petal.

SECT. IV. *On the Manifestation of Electric Currents in Fruits and Tubers.*

BECQUEREL[x], DONNE[y], and WARTMANN[z], have shewn, that when the electrodes are inserted into different parts of a fruit or tuber, effects upon the needle occur, amounting to 15° or 30° or more, depending upon the parts in which they are inserted and the kind of fruit. In some tubers, such as the potato, the beet-root, and in the carrot, the external layers were *positive* to the internal. In the *Tropæolum tuberosum* and the *Ullucus tuberosus*, the effects were inverse.

Some experiments were made upon the vegetable marrow and cucumber whilst attached to the plant. The electrode in contact with the external surface was slightly *positive* to the other when inserted into the centre. Similar experiments were made upon the apple, pear, and plum, when attached to the tree, or fresh gathered, or some time after they had been gathered. The greatest effect appeared with some

[x] In his second Memoir.
[y] Traité de l'Electricité, tom. iv. p. 164.
[z] *Loc. cit.*

of the apples, depending, however, upon the parts in which the electrodes were placed. No definite effects could be obtained; and as the effects might be referred to secondary actions, I do not think it necessary to particularize the results. BECQUEREL[a], in speaking of some similar results obtained by DONNE, adds: "Les courants ne doivent pas être attribués à la présence d'un acide et d'un alcali, mais bien à l'hétérogènéité des parties constituantes des fruits."

In the potato, carrot, and beet-root, similar effects to those observed by BECQUEREL were obtained. In radishes the external surface was *positive* to the centre.

The following general conclusions may be deduced from the foregoing experiments:

1st, That when the electrodes of a galvanometer are brought into contact, one with the surface of the leaf, and the other with the sap flowing from the same leaf, an effect occurs upon the needle indicating the surface and the sap to be in opposite electric states. These effects cannot be referred entirely to ordinary electro-chemical actions, but may be referred, in part, to the organic changes which take place in the leaf during vegetation.

2nd, When the electrodes are brought into contact, one with the external surface of the spongioles of a plant, and the other with the sap ascending from the root, the sap and the external surface are in opposite electric states. The effects which are here

* Traité de l'Electricité, tom. iv. p. 164.

observed with the galvanometer may, in the majority of instances, be due to ordinary electro-chemical actions, but in some instances the effect cannot be referred to these actions, but may be referred to the organic changes which occur in the roots during vegetation.

3d, That with the petals of flowers slight currents were obtained; and,

4th, In fruits and tubers powerful currents may be occasionally obtained, but these effects are evidently *secondary* results, due to the reaction of the different vegetable juices upon each other.

CHAP. XIII.

I SHALL now briefly recapitulate some of the conclusions that have been deduced from the experiments and reasonings detailed in the previous chapters. Sufficient evidence I think has been brought forward, shewing, that the actions which take place in the living body, in plants as well as in animals, viz. *secretion*, *absorption*, and *nutrition*, by whatever term they are designated, whether vital or organic, are accompanied with the manifestation of *current force*, and therefore present the marked characteristics of *polarity;* and consequently, the *force* associated with these actions, whether it be called *organic*-force, or *cell*-force, or *germ*-force, or by any other term, must of necessity be a POLAR force. It will be as well perhaps to avoid using the term vital force in these discussions, inasmuch as it is too general in its application, embracing as it does the whole class of vital phenomena, including those connected with the mind. But to the term *organic* I see no objection.

The class of actions which present the greatest resemblance to those which take place in the animal body, as well as in plants, are undoubtedly those of *osmose;* the similarity is too great to be overlooked. *Secretion* and *nutrition* may be considered as instances of *organic exosmose*, whilst *absorption* (*lacteal* and *lymphatic*) may be considered as cases of *organic endosmose;* the prefix *organic* will serve to mark sufficiently their differential characters. My object, however, is not to enter so much into a minute comparison of all these actions, as to point out their general characters, and more especially their marked characteristics; to shew, in short, that they are *polar* phenomena; and I shall now proceed to detail some of the consequential results which may be deduced from these investigations. And, *First,* in regard to

The Blood. We find that during *secretion* and *nutrition* the blood is in an opposite electrical state to that of the secreted product, as well as to that of the muscular or nervous tissue; it is in a *positive* state, and this electrical condition is produced and maintained by these actions. We find no difference between the arterial and venous blood in regard to their electrical properties, they both present the same *positive* state. Possessing this peculiar electrical character, there can be no doubt that a great many of the phenomena associated with the blood, and manifested during its circulation in the animal body, and which have been termed vital, are referable to this electrical condition. A further knowledge of this electrical state of the blood is requisite both to the physiologist as well as to the pathologist.

Secondly. In regard to the secretions and the

tissues. From the experiments of VASSALI EANDI it would appear, that the secreted products retain their peculiar electric state (the *negative*) when separated from the body. I have no experiments of my own either to prove or disprove this supposition. But with regard to the tissues, the muscular and the nervous, there can be no doubt that they do retain their peculiar electrical state, which is also *negative*, like that of a secreted product. Now this electrical state of the tissue would appear to be intimately connected with its vital condition; that the so-called *irritability* of the muscular tissue, as well as the so-called *sensibility* of the nervous tissue, are most probably entirely dependent upon it.

With respect to the *muscular tissue*, it would also appear as a necessary consequence of its electrical state, that the particles or substance of the tissue must be in a state of self-repulsion, in a polarized state or condition, a charged state, a state of tension, and that relaxation of the muscular fibre would be synonymous with this state of tension. Whatever would interfere with this electrical state would necessarily diminish this self-repellent action, and cause an approximation of the particles, in other words, a shortening of the fibre, and hence would arise muscular contraction. Now in favour of this supposition, and presenting a strong argument in confirmation of this view of muscular contraction, we are enabled to shew, that during muscular contraction electricity is evolved.

In regard to the *nervous tissue*, this can also be shewn to possess the same electrical characters as the muscular tissue; and this fact shews that they are evidently due to one and the same cause, viz. *nutrition*. Whatever remarks have been made in regard to the muscular tissue, may also be applied to the nervous tissue respecting its electrical condition; but I have not been enabled to satisfy my mind that any electric force is evolved in the nerve *during* nerve action, as is the case with the muscles *during* muscular contraction. Du Bois Reymond, however, states, that he has been enabled, under certain conditions, to obtain an *increase* as well as a *decrease* of the so-called *nerve current*. My experiments, on the other hand, shew a *decrease*, but no increase. The supposition that *nerve force* is similar to *current force*, is certainly not borne out by my own investigations; at the same time there is every reason for considering it as a *polar* force, from its intimate connection which may be shewn to exist with other forms of polar force, as in the electric fish, for example. The questions which physiologists will have to decide appear to me to be reduced to the two following: 1*st*, Is the electric condition of the nerve merely a condition necessary for the manifestation of nerve action? or, 2*ndly*, Is the electric force *converted* as it were into nerve force *during* nerve action? Now if the *decrease* in the nerve current, which I have observed to take place under these circumstances, be due to a *loss* of the

electric force, and not to a disorganization of the
tissue which I have supposed it to be, then we get
evidence of the *conversion*, as it were, of the electric
force *into* nerve force. We need not expect to have
any great effect upon, or any great loss in, the nerve
current, as an *equivalent* during this conversion. The
quantity of electricity associated with a grain of
water, and evolved during its decomposition, its
equivalent, when existing in the static form, as in a
charged Leyden jar, equals that of a violent thunder-
storm; and so in regard to nerve force the electricity
manifested as nerve current, a loss only of 3° or 4°
might be equivalent to its conversion into nerve
force, but then I think we have a right to expect,
that the *decrease* in the nerve current, to whatever
amount it might be, ought to be sudden and
definite.

The same arguments and the same reasons may
be applied to muscular action; the *loss* which
occurs in the muscular current *during* muscular
contraction may be accounted for, by the electricity
becoming free; and there is also a developement of
heat manifested at the same time; but in the nerves
there is no manifestation of any electricity becoming
free. The questions may arise, whether in the
action of nerve force upon muscular force it is the
conversion of nerve force into electric force that
becomes manifested, or whether it is the electric
force of the muscular tissue that is evolved. And in
regard to the fish, the same question may be put,

does the electric force exist as such in the organ, and is merely evolved under the influence of the will, or is nerve force *converted* during this action into electric force? Now with regard to the first question, I believe that it is the electricity of the muscular tissue that is evolved, for we can prove that the muscle possesses a previous electric state. But with regard to the fish, I am not aware that the substance of that organ has been proved to possess a similar electric condition. The muscular tissue possessing an electric condition, would go far to prove that nerve force must bear some intimate relation and connection with the electric force to cause its manifestation. In all these discussions, however, it is important to bear in mind, that in the exertion of force, whether muscular or nervous, there must be a loss of power, and as they are both dependent upon nutrition for their manifestation, so they are dependent upon nutrition for their restoration. NUTRITION (including under this head the kindred actions *secretion* and *absorption*) would appear to be the ultimate and most important act in organic life.

DR. CARPENTER[a], in a valuable Paper *On the Mutual Relations of the Vital and Physical Forces*, speaking of the term ' *germ force*,' as employed by MR. PAGET to designate the power which each germ possesses ' to develope itself into the perfection of an appropriate specific form,' adds, " so far from regarding the

[a] Philosophical Transactions, 1850, p. 727.

whole force which produces the evolution as being possessed by, or as residing in, the germ, it will be the author's object to prove that it is of *external* origin." . . . . " That the vital force which causes the primordial cell of the germ first to multiply itself, and then to develope itself into a complex and extensive organism, was not either originally locked up in that single cell, nor was it latent in the materials which are progressively assimilated by itself and its descendants; but is directly and immediately supplied by the Heat which is constantly operating upon it, and which is transformed into vital force by its passage through the organized fabric that manifests it." Now the influence of external agencies upon the developement of an organism is not to be denied; but when we speak of Heat being converted into vital force, or that · Light is converted into vital force, that Heat and Light are absorbed as it were and converted, we are then considering these *forces* as entities; not that Dr. CARPENTER goes to that extent, for he speaks of the necessity of a ' *material substratum;* ' but the language employed leads to this conclusion. The error arises from the employment of the term *conversion*, which expresses inadequately the meaning that is intended to be conveyed by it. When we speak of electricity being converted into magnetism when a current of electricity traverses a helix, the two forces, magnetism and the current, are produced and exist at the same time; there is no change of the electric current into magnetism. Conversion implies a change, a loss on one side and a gain on the other.

FARADAY[b] commences the ' *Nineteenth Series of his Researches*' with stating, "I have long held an opinion, almost amounting to conviction, in common I believe with many other lovers of natural knowledge, that the various forms under which the forces of matter are made manifest have one common origin; or, in other words, are so directly related and mutually dependent, that they are convertible, as it were, one into another, and possess equivalents of power in their action." DR. CARPENTER[c] observes, "starting with the abstract notion of Force, as emanating at once from the Divine Will, we might say that this force, operating through inorganic matter, manifests itself in electricity, magnetism, light, heat, chemical affinity, and mechanical motion; but that when directed through organized structures, it effects the operations of growth, developement, chemico-vital transformations, and the like; and is further meta-morphosed, through the instrumentality of the structures thus generated, into nervous agency and muscular power." It appears to me, that we should be justified in considering this Force which DR. CARPENTER speaks of to be POLAR FORCE, and that the different *forms* under which the various forces are made manifest, as electricity, magnetism,

b Experimental Researches, vol. iii. p. 1.

c *Loc. cit.*

heat, light, &c. are merely different manifestations of this one Polar Force, and therefore have a common origin. One great object experimentalists have to determine is the polar character of the phenomena, and to point out the connecting links which bind these different phenomena together.

From the facts recorded in the present Essay, we can come to no other conclusion than that *the force manifested as organic force in organized beings is a polar force.* The influences which these external forces, Heat and Light, have over organic developement is not so *direct* as would at first sight appear; their influence is not to be doubted, but they appear to be excited in inducing changes of a chemical nature. Heat and Light without moisture and food would have no influence over organic actions; and it is questionable whether there is any loss of heat or of light indicating a conversion as it were of those forces into organic force during the developement of the organism. The connection, the relation that exists between organic forces and physical forces is manifested by the polar character of the phenomena, by the manifestation of current force during organic actions; and perhaps it would be more correct to say, that the chemical force of inorganic bodies is converted during nutrition and absorption of food into organic force; and, consequently, the same influences may be expected to be exerted by Heat and by Light over organic actions, as are exerted over ordinary chemical actions.

There is one circumstance which has appeared to me deserving the attention of physiologists, and one that has generally been overlooked; I allude to the influence of magnetism over organic bodies. Living as we do in a constant field of magnetic force, and subject as all bodies are to the influence of terrestrial magnetism, there can be no doubt but that magnetism must exert some influence over organic actions. "When we remember," says FARADAY[d] in 1845, "that magnetic curves of a certain amount of force, and universal in their presence, are passing through these matters, and keeping them constantly in that state of tension, and therefore of action, which I hope successfully to have developed, we cannot doubt but that some great purpose of utility to the system, and to us its inhabitants, is thereby fulfilled, which now we shall have the pleasure of searching out." I have related elsewhere[e] some experiments on the Influence of Magnetism over Chemical Action, which were undertaken as a preliminary step to an inquiry into the influence of Magnetism over Vegetation; but the means at my disposal were totally inadequate to carry on the requisite investigation, and I was consequently compelled very reluctantly to abandon it. The inquiry may be considered by some as frivolous and useless, but to me it appears to be one deserving of consideration.

[d] Experimental Researches, vol. iii. p. 79.
[e] Edinburgh New Philosophical Journal, April, July, 1857.

www.ingramcontent.com/pod-product-compliance
Lightning Source LLC
Chambersburg PA
CBHW021801190326
41518CB00007B/392